DEEP ENOUGH

DEEP ENOUGH

A WORKING STIFF IN THE WESTERN MINE CAMPS

BY
FRANK A. CRAMPTON

FOREWORD BY W. H. HUTCHINSON

UNIVERSITY OF OKLAHOMA PRESS: NORMAN

Library of Congress Catalog Card No. 81–43639

TO

John T. Harrington and Michael Sullivan,
hard rock mining stiffs and loyal friends

TO

The prospectors and desert-rats who found the mines
and the mining engineers and hard-rock mining
stiffs who made them pay, and to
Old John Lamoigne, the embodiment of them all

TO

James D. Stewart, Mining Engineer, of Gold Run, California

AND

Charles F. Willis, Editor of the Mining Journal, *of Phoenix, Ariz.*
Men of understanding, principle, and courage to whose
work and untiring effort the mining fraternity owes much
I dedicate this volume in deep appreciation.

FOREWORD TO THE SECOND PRINTING

by W. H. Hutchinson

Even as did Huckleberry Finn and The Ancient Mariner, Frank Crampton made a splendid wayfaring and lived to tell about its first thirty years, 1888–1918. In the telling he opens wide a sizable picture window that affords a panoramic view of the closing years of the individualistic mining West, where Goldfield, Nevada, made its last "beautiful, bibulous Babylon." In the telling too he brings alive for the amazement of more pallid generations the self-portrait of a man who in the bright morning of his youth seized life with both hands and did his best to wring it dry. He came as close as mortal man can come to doing pre-damn-cisely that!

What he did and learned and made of himself in those years may seem incredible today. That what he writes of himself and his deeds rings true as a shod hoof on *malpaís* is attested by the letters he received, after this book made its original appearance, from desert rats and hard-rock stiffs who had known him or known of him in the times of which he writes. The truth of his account is attested as well by his original publisher, the late (alas, the word) Alan Swallow, who had a sure sense for what was real and what was false in writings about the West that was. It is attested beyond all doubt by the experiences of my own father, who hit the Colorado hard-rock camps while still in his teens and went on to Tonopah and Goldfield, Nevada, shortly before Crampton began serving his apprenticeship in life underground, and made a name for himself in much the same way that Crampton did. For these reasons it seems safe to say that you may depend upon what

Crampton writes, as upon the points and shards and tree-ring datings of a vanished people.

Born in New York City into a well-to-do, socially prominent, and even more socially conscious family (Norman Rockwell became his stepbrother), the experiences he had in the mines made him a card-carrying member of the Western Federation of Miners and a supporter of the more radical Wobblies, known formally as the Industrial Workers of the World. In his accounts of the mines and the times when "men were cheaper than timber," you can learn more about the appeal of those organizations to the men who worked "down the hole" than you can from the perfervid rhetoric of their intellectual enthusiasts today. His account of experiencing "cold-boiled flesh" during a ten-day entombment underground makes this point plain, as do his recollections of the Ludlow Massacre in Colorado, which he witnessed after trying vainly to prevent it.

He learned the art and craft of hand mining in hard rock well enough to make half the team that won the two-man rock-drilling championship in Goldfield the day before Tex Rickard staged the Gans-Nelson fight in that tumultuous town. In his seventieth year he alone drilled twenty-two inches in solid rock using a four-pound single jack, or hammer, just to show the modern machine miners how it once was done.

He learned how to "ride the rods" beneath freight cars, which had them then, and how to "ride the blind," meaning the open half of the telescoping vestibule of a passenger car coupled just behind the engine. He was wont to exercise these skills even when he had the money to "ride the plush." These practices of bindle stiffs, who, he held, were quite distinct from hoboes, were frowned upon by train crews, who were not averse to making the nonpaying rider quit the train immediately, no matter what its speed or the terrain at the moment. Crampton also taught himself assaying, mineralogy, geology, surveying, and the meanings of dips, spurs, and angles. In time he became an entrepreneur who made money in large gold and silver chunks and spent it and lost it with equal éclat.

He mined for uranium in southwestern Colorado before World War I and rescued Zane Grey from his own ineptitude in Death Valley. He knew and respected the Cornish miners' belief in the benign but unseen underground creatures whom they called tommy-knockers, who warned miners of impending danger; and he was always conscious of what the earth told him as it reacted against the onslaughts of puny mortals burrowing in its bowels and leaving "the deep shafts crying to be filled."

He knew Old John Lamoigne and buried him after he succumbed to the August heat of Death Valley—to which Old John had come from France as Jean Lamoigne in 1884, bringing with him volumes of classical literature in French, English, and German. Lamoigne was the dean of the desert prospectors, and in Crampton's opinion, Old John's "castle" on the fringe of Death Valley gave the castle idea to the more notorious Death Valley Scotty.

Crampton is kinder to Scotty in this book than he was in conversation. He had definite ideas about where Scotty got his funds, long before Scotty met the Chicago angel who, accepted doctrine holds, bankrolled his major extravagances. During the Goldfield boom Scotty had ample funds which, Crampton holds, were derived from buying stolen high-grade ore from the miners who purloined it. Goldfield's mines had ore that often ran more than $4 a pound, an irresistible temptation to men who were earning $4 per ten-hour shift mining it. Crampton also held that Scotty had another source of income, unreported in this book, from operating two whorehouses in Goldfield, named the Cottage and the Oriental.

Crampton's matter-of-fact treatment of whorehouses as an essential component of western mining communities may be a revelation to those whose conception of such "houses of ill fame" has been shaped by flamboyant media presentations. That they often provided more than just carnal ejaculation is made clear in this book's chapter on "The Girl from Goldfield." There is a twist to that story that would have delighted O. Henry.

My review of *Deep Enough* a quarter-century past brought me several visits from its author. During those visits, in the words that Rudyard Kipling gave Mahbub Ali-the-muleteer, "The talk slid north and the talk slid south, with the sliding puffs from the hookah-mouth." Crampton told how he had been forced to kill a man in Goldfield, who had held a grudge against the man from whom Crampton purchased his first assay office. The aggrieved one saw his enemy, as he thought, standing in the doorway of the assaying shack and unlimbered his artillery without perforating anything but the desert air and ground around the premises. He then reloaded and began firing from closer range, which induced his target to get the shotgun he kept against culinary emergencies (that is, for activities such as rabbit hunting) and put an end to the proceedings. The automatic verdict was self-defense.

He told me casually, as one might mention crossing a street to get to the other side, that between mining ventures he had served as an adviser to both Sun Yat-sen and Chiang Kai-shek, before service in the Air Force in the Pacific during World War II. More mining ventures were then followed by service as an adviser to Syngman Rhee shortly after Rhee assumed the presidency of Korea, and then again came mining ventures. At the time of our meetings Crampton was serving as a consultant to the California legislature while endeavoring to find the time and the capital to reopen a mine near Wickenburg, Arizona, from which he and a partner had taken a sizable amount of free-milling silver ore before 1918.

From these and other things that Crampton told me, and from what *Deep Enough* already had imprinted on my mind, I thought of him then, as I remember him today, with the words applied to the Elizabethan sea dog Sir John Hawkins: "Nothing would serve him but the wide world to walk in." No man of his generation and temperament could ask for a more fitting accolade and epitaph.

PREFACE

Deep Enough is an autobiographical account of some of the adventures that befell me and my companions, and of events in which I participated. The persons who participated are identified by their true names with but one or two exceptions.

The title, *Deep Enough,* expresses my attitude toward many happenings during the period that it covers. The expression "deep enough" probably originated in Cornwall and came to the United States with Cornish miners—"Cousin Jacks" they were usually called; their wives and women "Cousin Jennies." At first the term referred to a drill hole in which powder was placed for blasting. When the hole had reached a desired depth, it was "deep enough." Later the use of the term broadened to include anything one did not like or wanted nothing more to do with. However it was used, the attitude of "I don't care" was ever present. Today the equivalent would be "that's that" or "I've had it."

The pictures in *Deep Enough* are reproduced from a few recovered of a collection of several thousands, after the Glendale-Montrose flood, of New Year's Day, 1934, in the Los Angeles area. My home was completely destroyed by flood-water, and very little that had been in it was recovered. After almost two months of effort, three photographers, Lynn Blakely, Melvin Ellette, and L. A. McArthur, of Willows, California, produced pictures that could be used for reproduction. Many of the prints and negatives on which they worked were mud-covered, stuck together, stained, and, with but few exceptions, apparently worthless. I extend my deep apreciation to them for the excellent work they did.

The pictures of Mark Twain in a bathing suit and of the group in which he appears with President Cleveland and political figures and bankers of the day, are reproduced from original prints, also recovered after the Glendale-Montrose flood. I remember "snapping" the pictures after my father had set up and focused the "Detective" camera, a gift from Mathew Brady, the great War of the Rebellion photographer.

Looked back upon over a span of forty or fifty years, life then might now be considered colorful, but we did not think so at the time. Certainly there was nothing romantic about it. Life was cold, hard, and demanding; often there was danger with risks weighed, calculated, and taken. Events that at the time of their occurrence seemed comonplace (which they were, because they were things we encountered and must face in everyday living) may seem, to a generation almost half a century removed from our hardships, privations, and individualistic living, strange or incredible; and I hope that they will prove interesting.

Hard-bitten pioneers were still developing the "West" when I reached it in 1904. The atmosphere, color, and spirit of risk and adventure prevailed. Mining camps and boom towns, some of them established decades earlier, had not changed greatly from the life and traditions of former days. Goldfield, the greatest mining boom since 1849, Nevada's Comstock Lode, Alaska's Klondike, or Tonopah, was in its swaddling clothes.

Transportation was mostly by saddle animals, occasionally by buckboard, often afoot behind burros. Not infrequently, a trip of one hundred miles or more, sometimes several hundred miles, was made on foot. The old stage operators had almost given up. The few that had the hardihood to stick it out continued to operate Concord stages, which carried passengers and mail, but their days were numbered. The "horseless carriage" was an unusual sight except in the larger cities. The few daredevils who tried to negotiate desert roads with them wished that they hadn't.

Roads between mining camps and boom towns were the most traveled. They were nothing to write home about: either dusty or deep with mud during the wet weather seasons, which fortunately were not for any length of time. Many sections of the roads often resembled rollercoasters. Some had sections of log-corduroy over the long sandy stretches. If there was no corduroy, we used our bed-roll tarps to cover the sand. The process was nothing unusual, in fact something that we expected; but it was time consuming to move a tarp from behind a buckboard, and, in later years, a Model T, to ground it must pass over.

Water was always a problem. It was not unusual to encounter stretches of road with fifty to seventy miles between one spring or water hole and another. There were no houses along the way, nor were there filling stations or repair shops.

Roads away from those that were well traveled were mere marks on the desert landscape. It was possible sometimes to find them by close observation and by knowing where they should be and where they should go. Cloudbursts, with the resulting flash floods, frequently destroyed even these traces of earlier passages, but signs were there for those who had learned how to read them. Ruts of earlier wheel tracks became gullies cut by the rushing waters. We would have to detour until we found again the faint, but undisturbed, marks of the little used route. Getting around the washed-out sections was hard work, with pick and muck stick to smooth the way. Often we left all signs of roads behind and independently headed for wherever we were going, making our own new road as we progressed.

It was easier when the Model T came into our lives. We did not have to chain-hobble it at night and hunt for it in the morning as we had had to do with burros, mules, and horses. There were the same drawbacks, however, of carrying supplies: instead of food and water, however, it was fuel and water. In the Model T we carried the customary ten to twenty gallons of water, never less than ten gallons of gasoline and several quarts of oil. In addition, there were our regular bed rolls, clothing, food for us, and a complete machine shop for the car. There were spare parts to replace anything that might break, especially axles, drive shaft, rear-end gears, front wheel spindles, and what not, to say nothing about spare tires, tubes, and tire repair outfits. We might arrive at our destination sooner, but there was no discernible lessening of work.

There were few towns, all cut from the same pattern, and each was near a group of mines. They had narrow streets with a few stores, eating places run by Chinese who knew how to cook, sometimes a bank, not often an "opera house," but always saloons and gambling places. A few homes were built on narrow streets away from the main drag; their number was small and often the owners took in boarders, mostly hard-rock stiffs who worked at nearby mines. There was always the "line" or "stockade" with its dance halls, parlor houses, and cribs, where working stiffs, and business men of the town, too, went for relaxation and the companionship of the girls who lived there.

Life at the mines was tough. Mostly the stiffs lived in bunkhouses, often as many as thirty in a single building, with wood-bottom bunks, sometimes two to a tier. No bedding was furnished—each stiff had his own bedroll, or a bindle. A cookhouse usually furnished excellent chow for ravenous appetites. When reading became difficult by flickering candlelight or coal-oil lamps, the stiffs went to the saloons, gambling places, or the line, for there was nothing else to do.

Working conditions in the mines, wherever one went, were about the same. A mucker had to muck and tram sixteen one-ton cars in an eight-hour shift, or twenty if the shift were ten hours. Two cars an hour was the minimum wherever the stiff mucked and trammed. If there was a trammer to push the ore car, the job of the mucker increased, for it was continuous mucking to fill cars while the trammer took out a loaded car and brought back an empty. Hard-rock stiffs drilled single or double jack, and they drilled a round and shot it before tally, on every shift. In the larger mines there were sometimes "machine" drills which made it easier in hard rock, but hand drilling was the preferred method if the hard-rock stiff was a good one.

There were few air blowers, or suction fans, in use, and the stiffs worked in lung-choking, eye-smarting, powder gas for an hour or two after their shift went on. There was little, and, sometimes, no, timbering—men were cheaper and expendable. There were scores of stiffs waiting for a job, many of them bohunks, but all good. If a stiff didn't like the job and it was "deep enough," or if he did not "put in a shift," he went down the hill, with his time, talking to himself, and some other stiff, in just as hard luck, took over the job.

There has been a tendency to berate and degrade labor union activities of the days following the turn of the century. Some of it may have been deserved, but the I. W. W. of that time, and the Western Federation of Miners, did a fine job for the working stiffs. I would not now be around had they not done so. Mine owners were hard, calloused, almost ruthless, and organized. A working stiff did not have a chance to uphold his rights to anything. Blacklisting was not uncommon. Management of the larger operators thought little of anything but making money. There were no safety laws, workmen's compensation, or state and federal legislation that protected the working stiffs. If one were hurt, he was almost always "on his own." Sometimes, but not frequently, if a man was killed, a few dollars were given to his family, usually not even enough to bury him. For the stiff who was badly hurt, or disabled, it would have been better to have been killed in some cave-in or other accident, as were many.

I reached the "West" in 1904, but it was not until 1906 that I saw the Pacific Ocean for the first time. San Francisco and Los Angeles were in the traditional stage between the "old" and the "new," but tradition and atmosphere clung to them as a brilliant sun-pierced fog. The "foreigners" from east of the Mississippi could not destroy what was there. Nor did they wish to do so. They were soon willingly absorbed, and no one could recognize them as newcomers.

A few days before the San Francisco "fire" I shot ducks and a goose on marshy land near what is now called Lake Merced, within the city limits. Not long before that I was in Los Angeles where I shot rabbits on Wilshire Boulevard in beanfields owned by Matt Wolfskill, whose ranch house stood near Beverly Glenn, not far from the Sawtelle Soldier's Home. On both sides of Wilshire Boulevard were vast oil fields, oil well derricks extended as far west as Beverly Hills, and east to Westlake Park.

At that time, December, 1905, Charlie Waring and I found what we thought were fossil bones near an oil pool west of what is now La Brea Boulevard. These we took to what was known as the Los Angeles Chamber of Oil and Mines, and showed them to J. Nelson Nevius, W. W. Orcutt, and others. We told them the bones were being crushed and used for chicken feed. The following day we went to the oil pool with Nevius and Orcutt. Not long afterwards we heard that the University of California had men working at the pool, now Hancock Park, uncovering a wealth of fossil bones and geological information.

In the fall of 1951 I made an inspection of Sixth Army installations with Colonel M. J. ("Mike") Blew, Corps of Engineers, USA, retired, who had come from Washington, D. C., to accompany me. During the trip through California, Arizona, and Nevada, we often left paved highways and drove from one installation to another on old dirt roads through country I had often traveled, in former years, behind burros, in a buckboard, or in a Model T.

As we went on our way, I told Mike of things that had happened in the places we passed or visited and in other places as well. Not a few old mining prospects, and mines also, had been taken over, and great areas surrounding them, by the National Park Service, or the United States Army or Navy, for training troops and personnel.

When we reached Camp Irwin, one of the important stations on our tour of inspection, I told Mike that my brother Ted and I once had established a camp not far from where Post Headquarters now stood. At the time, we were operating several silver prospects nearby, and the camp we had established was central to them all. We had no serious camp problems, excepting that of water, which we were hauling from Paradise Spring some miles west, and from Garlic Spring, three miles south. The number of men we had working for us was increasing, which made water hauling uneconomical; we either had to get water where we were or move camp to water, which would be inconvenient.

When we had studied the geology of the area, we decided that the granite underneath our camp was an overthrust, probably covering an ancient lake bottom, or river bed, in which there would be water. We sank a well shaft through the granite and, as I remember, at somewhere about two hundred feet we went into the gravel we had expected, where we found more water than we had thought possible. When we abandoned our mining operations at that locality, some years later, I proceeded to forget the mines and the water as well. When the Second World War began, the Army Engineers looked for a site for a training center and a bomb-range where there was water. Our old well gave the Army Engineers the clue to the water they needed. They established Camp Irwin and put down other wells, not by single jack, windlass and ore bucket, as we had done, but with modern drilling outfits. Today these wells provide all the water that is needed for thousands of troops in training.

In leaving Camp Irwin I took the old road to Yermo in order to stop at Garlic Spring. I had saved the story of Old John Lamoigne—christened Jean Francois de Lamoignon, when born in February, 1857, at Lamoignon, France, educated in England, graduated from Ecole des Mines, Paris, and student at Heidelberg—who had come to Death Valley to try to make his fortune in borax.

Sitting under a scrawny tree that had grown up since Old John's time, I told Mike his story, and of the "castle" Old John had built, and of Death Valley Scotty, who took the idea of a castle from him and built one of his own near Death Valley, years after Old John had passed away. I spoke of Death Valley and Furnace Creek Ranch, at the mouth of Furnace Creek Wash, where Oscar Denton used to hold forth as foreman for the Pacific Coast Borax Company, which operated it in connection with their borax operations in and near Death Valley.

I told Mike how the Furnace Creek Ranch in those days was the place to which one and all of us went to rest, and to clean up in the warm, soft water flowing in the ditches designed, from springs in the wash, to furnish irrigation water to the fields of alfalfa on which the ranch's cattle, horses, and mules grazed. Not only was the washing and cleaning-up process a welcome one, but also bedrolls could be spread and rest taken under spreading shade trees in green fields. Of equal importance was the fact that food supplies could be ordered through Oscar Denton to be brought to the ranch from Death Valley Junction. Waiting for the supplies was a most welcome respite.

There were other advantages of camping at Furnace Creek Ranch. Necessity required Oscar to butcher a steer or two, always at night when the air was cool. A wagon would come down from the Lila C, and in later years from Ryan, to take a couple of the butchered steers to the cookhouses for the borax mines. Oscar always had fresh beef on hand, which was a most welcome change to the usual bean, bacon, and flapjack diet of the prospector.

It was not often that more than one or two of the old timers, prospectors, or desert rats were at the Furnace Creek Ranch at the same time, but Oscar was a good listener, and he passed the latest stories and the best lies to the next to come for the regular clean-up and supplies. The Furnace Creek Ranch was, therefore, the place to which everyone went whenever loneliness overcame him and he needed human association and conversation. The old time Death Valley prospectors traveled alone, their burros the only companionship they had. Without Furnace Creek Ranch, Oscar Denton, and the Panamint Indians, Death Valley would have been intolerable.

Occasionally we would go to Greenwater or Ballarat, rarely to Skidoo. In duck season we went sometimes to the Indian Ranch near the mouth of Wildrose canyon, frequently to Shoshone where Dad Fairbanks and Charlie Brown helped to make our visit worth while. The old timers and prospectors who lived anywhere near Death Valley went, mostly, to Furnace Creek Ranch, because of its central location, trees, green meadows, fresh clear water, and Oscar Denton.

I also told Mike about the song we composed, to which we added new verses whenever a few of us would meet. The last verses were made up by a large group, all of us prospectors, desert rats, and hard-rock stiffs, at the almost deserted camp of Greenwater in the fall of 1919, when most of the group got together. It was to be the last reunion for most of us, although we did not know it at the time.

Many of the verses have long since escaped my memory. Possibly a revisit to Greenwater and the ghosts of the past, and some high-power home-made whiskey, would help to revive them. The few verses that I do remember, however, reveal the depth of feeling against the hardships each old timer faced, and shared sometimes with another of his kind, on the desert, but which were accepted without comment or complaint as part of the chosen life. Other than in the verses of this song, I have never heard hardships mentioned by any of the old timers excepting as something to joke and laugh about, because what they did and what happened to them was commonplace. Had the old timers, prospectors, and desert rats not had to fight for everything they got, and then fight to keep what they had found and to make it worth while, they would have called it "deep enough" and let someone else take on the job. To them nothing was worth while unless it was earned the hard way.

When I had finished the story of Old John, and had told him about Death Valley, Mike insisted that I write about what I had told him, and of life in the mining camps and boom towns. The opportunity to write as Mike suggested did not come until 1955, after I had returned from Korea, where I had been attached to the office of President Syngman Rhee.

It has been interesting writing, for events long filed away in my memory have been relived, as if only yesterday. Pals, long gone but never to be forgotten, my brother Ted, John T, Sully, Jack Commerford, Old John Lamoigne, Shorty Harris, Pete Buol, and a host of others, have frequently walked beside me in my writing. It was good to have them close to me again, and to relive the days when we were together.

FRANK A. CRAMPTON

WILLOWS, CALIFORNIA
SEPTEMBER 1, 1956

CONTENTS

ILLUSTRATIONS

MAPS

DEEP ENOUGH

PRELUDE

My birth certificate reveals that I was born on Park Avenue, in New York City, on February 29th, 1888. The fact that I was a leap-year-day arrival did not seem to impress anyone, and least of all those closely connected with the event. Conversations that touched on the subject were unceremoniously sidetracked in favor of those about the great blizzard, which had been raging that day, and the inconveniences it had caused.

I am sure that many things that I did in my youth were to call attention to the fact that, after all, I did exist. In later years events simply happened without my precipitating them. I certainly gave my family enough to talk about, and, in view of some members, something to hide in the closet with other family skeletons.

My attitude in childhood was one of rebellion. I was dressed in Lord Fauntleroy costumes, kilts, and even dresses with lace. The greatest indignity was to wear curls, meticulously fashioned by a wonderful colored mammy, born in slavery, Rosa Artis—"Artie" I called her. Those curls were worked over, not just once a day, but whenever they gave signs of becoming disheveled. No ordeal could have been more painful.

I am sure that Artie and my paternal grandfather were responsible for my viewing life in reality. Artie and I had long talks about the outside world, about her life in slavery, where and how she had lived, of her slave father and mother, and of the Great Rebellion that made her free. My grandfather, who had been a surgeon in the Union Army and had amputated General O. O. Howard's arm at Fair Oaks, often joined in those conversations. He too was deeply concerned with fundamentals and reality.

My fourth birthday was a gala event for everyone but me. The previous day newspaper reporters had come to our house to talk to my mother and me. Mother had a picture of me taken by Dana, a week or two before my birthday, which I did not like. My costume was a long plaid dress to below my knees, and my hair was done in perfect curls. The reporters were given copies of this outrage, and it appeared in New York newspapers on Leap Year Day. (I lived that one down long ago.)

Two hours before I was to welcome guests to my birthday party I cut off some of my curls. Artie caught me before I could demolish them all. By some superhuman connivance with the devil, those cut-off curls were wrapped around something, and attached to their stems with glue. I did not enjoy the birthday party at all. The next day the world became much brighter, for Artie finished the cutting job and my curls disappeared forever.

Mother's maternal great uncle was Francis A. Palmer, president of the Broadway Savings Bank and the Broadway National Bank. He had gained a reputation as a philanthropist in endowing schools, colleges, and churches. He used to boast that he had started as a conductor on the Ninth Street Horse-Car-Line, and had saved nickels. After he had gained control of the Fifth Avenue Stage Coach Line, he became interested in banking. Palmer had been active in New York politics during Boss Bill Tweed's time, and was at one time City Chamberlain (1871-1872). He helped bring men from Italy and Southeastern Europe to the United States. When the immigrants arrived, Palmer secured jobs for them in the coal mines of Pennsylvania, Utah, and Colorado, or with contractors, railways, or any employer willing to pay him a fee for providing cheap labor. "Bohunks" the immigrants were called.

Sometime in my early childhood Mother received the gift of a house on Seventy-first Street, Central Park West, from this great uncle, Francis A. Palmer. There she entertained frequently. Among her guests were friends from her days at Wells College, including Frances Folsom, who married Grover Cleveland, and Elizabeth Burling, wife of Admiral W. T. Sampson, who later commanded the American fleet in the naval battle of Santiago, Cuba. I also remember General O. O. Howard, General Joe Wheeler, Joseph "Uncle Joe" Cannon, Chauncy Depew, Mark

Hanna, Dan Lamont, Richard Olney, and Samuel Clemens—"Mark Twain."

The Clevelands had asked mother to introduce Mark Twain to "Uncle Frank," which she had done. Mark Twain was badly in need of money at the time, and the Clevelands wanted to help him as much as they could and thought that if Twain could have the opportunity of having an informal and leisurely talk with Palmer something might be arranged to relieve his financial embarrassment. Mother managed to arrange the opportunity that the Clevelands wanted, and "Uncle Frank" invited the Clevelands, Stewart, Twain, and others to be his house guests at Craigville, Cape Cod, Massachusetts, where he was going for the summer.

When he extended the invitation, "Uncle Frank" told Mark Twain that he did not like his flowing white hair, and suggested that he have a barber trim and dye it. Twain showed up at Cape Cod with his hair slightly trimmed, dyed brown, and combed as tightly to his head as he could get it. "Uncle Frank" did not like the changed appearance of Twain, and told him to get the dye off. Mark Twain thereupon donned a bathing suit and took a few dips in the surf with my mother, while father, an avid camera enthusiast, took pictures. Twain's hair was restored to its natural whiteness, and the Clevelands and Stewart had a good laugh. "Uncle Frank," however, was annoyed because he did not approve of women bathing in the surf, although attired in bathing skirts, bloomers, and stockings, and chaperoned by a husband.

Time passed quickly during the years I was completing primary and high school studies in a military academy. During that time I showed an independence that was a great burden to my family. Some of my escapades were deliberate, others purely accidental and directed by a natural chain of events. Often I suffered according to the precept of the times, "spare the rod and spoil the angel." Angel was a misnomer, but I lived through it, and it made little difference. I was never impressed, though my rear hurt a lot, sometimes.

Money meant nothing to me. I was allowed to purchase toys and books at stores where my mother traded, and charge them to her account. The ten pennies I received as a weekly allowance was the only money I was permitted to handle. I learned more one summer, however, when I was given five dollars and sent into town, Englewood, New Jersey, where we were visiting the Kennedys, to make some small purchases for my mother. Roderick Kennedy—whose father Roderick J. Kennedy, who had passed his name to his son, was important in Tammany Hall and a gallant figure in Saint Patrick Day parades—was with me.

After we had finished shopping, guided by a list mother had given me, we were to buy ourselves ice-cream sodas, then go home. We did the shopping and had our ice-cream sodas. There was money left, over three dollars; so we bought candy, then sodas until we could hold no more. On the first leg of our way home we passed a shop, and we each bought a bow and some arrows. There was still a lot of money left. We had a problem—how to dispose of what remained.

Crossing a bridge over a stream on the next leg of our journey, we stopped and discussed the money situation as we watched the water flow under the bridge. Money was of no use if there was nothing more we wished to buy. The solution was to throw it into the creek. The coins zig-zagged until they rested on bottom, shining. We watched the two bills float downstream until they were out of sight.

When we reached home, the purchases we had made for mother were delivered to her. We were astonished when asked for change, and made to explain why there was none, while a silence hung over the room like a thundercloud. When I had finished, not a word was spoken. We were marched upstairs and the back of a hairbrush was applied to bare skin. Then a lecture about money followed, and an hour in a closet with the door locked.

The lesson on the importance of money never took root. In telling about the incident, when I was not supposed to be within hearing distance, mother would end her story by saying: "And you would never guess what they did with the change. They threw it into the creek!" Peals of laughter would follow.

When I was eight, my father died. A few years later my mother married Samuel D. Rockwell, son of a coal company executive of "Uncle Frank's" acquaintance. I inherited a step-brother, Norman, who at that time was more interested in a set of carpenter tools that I gave him one Christmas than in art.

Mother's marriage, when revealed, had repercussions that made headlines in the New York papers. "Uncle Frank" disinherited his grand-niece for "forsaking widowhood" as the New York *World* put it. The disinheritance followed disclosure of a harsh fact of life to Mother. After my father's death, "Uncle Frank" had persuaded Mother to endorse stocks and securities to him, so that she "would not have to bother with business matters." The disinheritance followed a request of my mother that the securities be returned to her. It was not until several years after "Uncle Frank's" death, and the subsequent general litigation initiated by other Palmer heirs, that my mother recovered a small fraction of what had been hers.

About this time, I had my first brush with the law. There were several of us on the grounds of the American Museum of Natural History. There was no sign warning not to pick flowers, as there was in Central Park, so my companions and I decided to pick some lilacs. When the law suddenly intruded, my companions escaped. I was less fortunate and was caught and marched home by a policeman.

It is probable that, had it not been about six o'clock and time for the policeman to be going off-duty, a simple warning would have been given and the matter forgotten. The "One of the Finest" who took me in tow no doubt had ideas.

When we reached home, I was marched inside. In no time there was a roomful of family, tears, hysterics, and smelling salts. Arrested by a policeman! How could I have done such an awful thing? The shame of it! The family was disgraced! The policeman was sympathetic, but firm. He must do his duty. To the police station I must go!

As the policeman took me in hand and pushed towards the front door headed for the klink, a fast thinking, and possibly experienced, male took a greenback from his wallet. What had appeared a terrible and incurable situation had found a remedy. The law departed, and the family regained its composure, and a hairbrush was brought into action.

After that, the law seemed to single me out as a confirmed and incorrigible criminal. There appeared to be two policemen assigned to watch my every movement. At least only two appeared to take any interest. One covered the parks and the other my activities outside of them. No matter what it was, or how many other companionable renegades were with me in some mischief, one or the other of the two police tormentors was always nearby to single me out, and let the others go scot free.

Small matters such as taking eggs from the nests of ducks along the shores of Central Park lakes, climbing trees, looking into squirrel nests, or climbing the cedar-post arbors, to pick wisteria, at the Seventy-Second Street entrance to the park, seemed always to end by a march home, escorted by the policeman who handled my park activities. The formula became routine, a tearful family, a duty-possessed policeman, something passing from hand to hand, the law departing, and myself awarded special disciplinary action.

The second policeman took care of everything outside of the parks. Rides on the steps of horse-drawn street-cars on Eighth Avenue, west of the park, were suddenly terminated. Removal from the rear steps of ice-wagons while I was collecting chips of ice on hot days, was not unusual. Annoying drunks who came out of the beer saloon on Seventy-First Street, opposite the Hotel Majestic, was a favorite pastime, and a prolific source of opportunities for my tormentor.

Once in a while, when our activities were more confined, such as playing cops-and-robbers, or pirates, on some vacant lot where "No Trespassing" signs were posted, the outside-the-park policeman would show up. There was not much deviation in method or routine. Those two policemen had it well worked out, their captures were made alternately. Share and share alike. It was a swell sideline, and they were scrupulously efficient in their police duties.

My family took a dim view of things. Not of the police activities, they were the law and infallible; but of the things I did to bring me into its firm and determined hands. Hairbrush or razor strap, with sometimes week-ends in bed, did not deter me or alter my views of independence of action.

By a twist of reverse fate, the Hotel Windsor fire on Saint Patrick's Day, a Friday, 1899, was the means of speeding my knowledge of mathematics. One of my cronies, Carl Patterson, another renegade, and I decided to see the Saint Patrick Day parade as it marched up Fifth Avenue with Roderick Kennedy's father in its front ranks, although we had been denied permission to do this.

We had better than a grandstand seat, on a window ledge of a house directly across the street from the Windsor Hotel. No one paid any attention to us during the parade, and after the fire broke out in the hotel, everyone was too busy with other things. We were overjoyed when the fire broke out, but wished it had waited until the parade was over. We had a double performance if there ever was one, although people jumping from the hotel windows made us squirmish.

I returned home that evening, long after supper, which my family had not eaten and had allowed to grow stone cold. Everyone thought I had been kidnapped, that it was another Charlie Ross affair. Every policeman in New York City was on the lookout for us and our abductors. All fear disappeared when I crossed the threshold. In its place was sudden resentment for my having caused so much worry and unpleasantness. I was escorted into the parlor by an ear and told to explain. I did, but it didn't go over.

My clothes reeked with smoke, and it took no time flat before the truth was wormed out of me. My mouth was thoroughly washed with soap. There were no half measures that night. A fine spanking with the back of a hairbrush followed, whereupon I was marched upstairs, disrobed, fed a bowl of bread and milk, given another chastisement on an already sore rear, and put to bed for a school week, with a week-end at the beginning and the end. I didn't worry about Carl, I had troubles enough of my own.

The next day I pondered my sin and found myself pure. Then came the problem of what to do, for nine days, with shades down. I dared not get up and play: someone down stairs would hear me, and when they did there would be chastisement on a rear that could take no more. There was nothing to do, or so I thought.

On the ceiling were myriads of small decorative flowers and leaves. I divided the decorations into patterns, then into numerical sequences, and for days my only amusement was in solving the problems in my arithmetic book for months ahead. Thereafter in the classroom mathematics was simple, and I fooled my teachers into thinking I was a good student. Years later I discovered that I had stumbled on the principle of the slide rule. I highly recommend that every youth do something important, important enough to be sent to bed for nine days, in a room with a ceiling profusely decorated. It would pay big dividends.

I also marched up Fifth Avenue or Broadway, often after dark and carrying a torch, in Gold Bug parades, with others from the military academy, joyously egged on by the elders with whom we marched.

My family was never informed, beforehand, of my political activities. When they learned about the episodes, they did not like it a little bit. No punishment was meted out, however; no doubt they thought that punishment would make me a Democrat, Heaven forbid! Those appearances in the parades were to show the world that I was against Bryan and free silver, and for McKinley, hard money, and the full dinner-pail. McKinley won, in my mind at that time, because I had marched in the parades.

The night of the election when McKinley won, with other renegades who had collected wooden boxes for weeks, we built a bonfire near the corner of Columbus Avenue and Seventieth Street. The street was paved with asphalt, one of the few in New York, and the fire was helped no little by this underlying fuel.

Alarmed occupants of nearby houses called the fire department, and the police. Being absorbed in feeding the fire, not one of the participants in the celebration knew of the impending doom, or had a chance to escape it. All fell into the hands of the law and were marched home by one of the New York City's finest. My captor, the policeman who watched my outside-the-park activities, was delighted. He was going to show Republicans that they must obey the laws of Democratic New York.

Word of my capture reached home, only a block away, before the policeman and I reached the front door. A reception committee of no mean proportions was waiting. The house was full of Republicans come to celebrate the election of McKinley. As the officer shoved me through the front doorway, he made some disparaging remarks about Republicans, their guttersnipe offspring, and the brats that built fires on city streets. He was laying the foundation and cornerstone for a real shakedown. Anything that had been handed to him on previous occasions would be peanuts.

The officer had no idea there would be a Democrat in a group of celebrating Republicans. There was, Roderick J. Kennedy's father, as staunch a Democrat as ever lived, and a political power in New York, regardless of who won a presidential election. Not that Kennedy had a victory to celebrate; he had come because he liked the Republicans who were in on the celebration, as well as celebrations.

Kennedy stepped forward and gave my captor a tongue lashing that I would still be proud to duplicate, on occasion. The officer knew who was bawling him out, I knew that when his hand let go of my collar and dropped to his side. When Kennedy had finished, the officer did not say a word, he just turned around and went through the front door, without anything paid for my freedom.

I never saw that officer again, and I learned later that he had been transferred to an outlying district, the Bronx. Neither did we again see the park officer who had given me so much personal attention. Kennedy was told, during the celebration that night, of their activities in behalf of law and order and the reduction of the family pocketbook. When we learned, later, that the park policeman was on duty in the Van Courtland Park district, life in the area of Seventy-First Street and Central Park West became worth living again. The ducks and birds in general had to conceal their eggs more carefully thereafter. The squirrels went unmolested, but the trees, lilacs, and wisteria arbors—and drunks from the corner saloon—caught hell.

The last four years of military academy afforded excellent opportunities for instigating untoward events. The headmaster, a hard-boiled colonel, veteran of the War of the Rebellion, with discipline running out of his ears, was in constant condition of grave concern over my conduct. The times I was caught were usually before some prom, or other important function, when my sweetheart of the moment was to be my guest. Confinement to quarters was so inevitable that my invitations were accepted only if I agreed to curb my propensities for receiving disciplinary action, at least until the event was over.

During the last four years at the academy I escaped a fate worse than confinement, several times. Hallowe'en of senior year was my big day, or rather night. We removed a heavy bell from the belfry of the administration building, and deposited it in the dean's office. For the bell we substituted a buggy, with a horse beside it, harnessed and ready to go. Had the belfry been longer, the horse would have been hitched between the buggy shafts. It was a major feat in engineering, and it had taken weeks of study, planning, and preparation to make it a success. No effort was spared to unearth the culprits, but none was caught.

Tradition declared that the senior class should plant a commemorative tree during the first hours of Arbor Day, between midnight and dawn. During the three years before I became a senior no tree was planted as tradition required. The ceremonies had to be performed in broad daylight, before the assembled cadets, an ignominy of no mean proportions.

Underclass organization had been perfect, and Clausewitz would have been proud to have included our tactics in his "Vom Kriege." In junior year the attack was planned so carefully that no tree was planted that night, and not a senior responded to reveille in the morning, nor was there one in chapel. They were in the cellar of the gym, bound hand and foot, with doors securely fastened to make internment doubly certain. There was not an underclassman missing in chapel. It had been a wonderful night.

Senior year ended in a blaze of glory. We planted our tree directly in front of the entrance to Administration at night. Deception, surprise, and a feint that diverted the rear guard at a critical moment, had accomplished our victory, made doubly gratifying by our planting the tree in the long-fought over, coveted location—the Administration flower bed.

Graduation exercises were over by noon of the big day. All that remained was the final review in the Armory of the 22nd Regiment, New York National Guard, which was to be followed by a dance and buffet supper. Our families, their friends, and our sweethearts, all were there.

The review came off in fine shape, then we retired to rooms in the basement to change from dress uniforms to fatigues. The remainder of the show was given by the graduated class. Back on the floor of the armory we fought the battle of San Juan Hill, with blanks in our muskets and gatling guns. It was a good show and went over big.

After the battle we again retired to the basement to change to dress uniforms, for the Seniors' farewell to the academy, and to turn over the responsibilities of the upper class to the Juniors. Our headmaster, the War of the Rebellion Colonel, had evolved a maneuver that was impressive. Returning to the armory hall we marched to the center where the Juniors were waiting at present arms. The formality of returning the salute over, we deployed around the Juniors and formed a hollow square, faced the guests, who were seated on all sides of the hall, and came to attention.

More than seventy of us were prepared to fire a parting salute to the academy, the incoming Seniors, and our friends. Orders were slow in coming, "ready" (we had already loaded in the basement), "aim," then "fire." It was a perfect volley, and from the mouths of muskets came garter snakes, falling into the groups of onlookers on all sides. All hell broke loose, but like the true soldiers we were, we ignored the tumult and confusion on all sides, went into company front, passed before the new Seniors, took their salute, and marched from the floor.

It had taken weeks to collect the small garter snakes, but in a few minutes, while changing into dress uniforms, the few of us in on the thing, put them heads down into every musket barrel, already loaded with blanks and in readiness to receive their unprecedented charge.

It couldn't be done now. The bore of rifle barrels is too small, but muskets took them in good shape. It had been a masterpiece. Nothing could be done about it, confinement to quarters or anything else, for we had been graduated. I was ready to face the world.

THE COMMERCIAL ADVERTISER, MONDAY, FEBRUARY 29, 1892.

HIS FIRST BIRTHDAY.

He Is Four Years Old To-Day, but Is Short on Anniversaries.

Master Francis Asbury Crampton, of No. 259 West Thirty-fourth street, is celebrating to-day the first anniversary of his birthday.

Master Crampton commenced his enjoyment early and had a royal good time this morning riding a bicycle around the floor of the nursery and banging a drum as he marched up and down and singing with all the strength of his young lungs. About

FRANCIS A. CRAMPTON.

noon his nurse dressed Master Francis in a brand new velvet jacket and a pretty silk kilt, and he went down into the parlor to receive with the assistance of his mamma forty or more young masters and misses, who came to take lunch and spend the afternoon.

A birthday is a rather important event in a person's life, but Master Crampton's birthday is many more times important than the anniversary of the average nat[...] day. He has so few of them. Every 29 [...]

A RALLYING POINT.

Republicans to Establish Permanent Headquarters Here.

HARMONY [...] BIG H.

WEBSTER IS COOL.

The Prisoner Still Maintains an Impassive Demeanor.

LAWYERS INDULGE IN A WRANGLE

Master Francis A. Crampton

requests the pleasure of your company

at luncheon

on the first anniversary of his birthday

Monday, February twenty-ninth

at one o'clock

259 West Thirty-fourth Street

New York

R.S.V.P.

AN INFAMOUS BILL.

The New York Commercial Advertiser *of February 29, 1892.* Crampton Collection.

THE EVENING JOURNAL'S CIRCULATION IS GREATER THAN THE COMBINED CIRCULATION OF ALL THE OTHER

THE 20TH CENTURY NEWSPAPER.

THE WEATHER—Generally fair and freezing to-night and Wednesday. light northwest winds.

NEW YORK JOURNAL

14 PAGES. W. R. HEARST 14 PAGES.

NO. 6,949—P. M. TUESDAY. NEW YORK, NOVEMBER 26, 1901. TUESDAY.

WED FOR THE SECOND TIME IN SECRET NIECE OF F. A. PALMER LOSES FORTUNE!

AGED MILLIONAIRE AND

Millionaire Philanthropist and Bank President Discovers His Selected Heir Married Again and Decides to Leave His Millions to Charity and Educational Institutions—Niece Kept Marriage Secret Three Years, and Is "Happy, Though Married."

It was reported to-day among

Headlines from the front page of the New York Journal, *November 26, 1901, announcing "disinheritance" of author's mother by her great uncle, Francis A. Palmer.* Crampton Collection.

House guests of F. A. Palmer (center, top row) and Mrs. J. Miller Crampton, the author's mother (on steps in front of Palmer). Beside Mrs. Crampton is Mrs. Grover Cleveland (in white shirt waist), who, when Frances Folsom at Wells College, was Mrs. Crampton's (Susie Frances Lewis's) roommate. Lower left, with hat in his hand, is Mark Twain (Samuel L. Clemens). Top row are, fourth from right, President Grover Cleveland and, third from right, Senator John M. Palmer. In front of Senator Palmer, on steps, is Arthur Sewall. Top right, sitting, is R. J. Kennedy; lower right, Dr. H. E. Crampton, the author's grandfather. Photo by the author's father. Crampton Collection.

Mark Twain (Samuel L. Clemens) at Craigville, Cape Cod, Massachusetts, summer, 1894. Photo by the author's father. Crampton Collection.

APPOINTMENT WITH THE WEST

The prelude to my engineering career began in the jungles back of the railroad yards in Chicago. My practical education started on a November day when, as a lad of sixteen, busted from an old ivy-league college, broke after running away from home and a family that thought I had disgraced it forever, I was taken in tow by two hard-rock mining stiffs and shown the ropes.

I had graduated from the military academy in the late spring, played around during the summer, and then headed off to absorb higher learning. It had been decided for me that I was to become a doctor, whether or no. My own mind was not made up as to what I wanted to be, a lawyer possibly, anything but a doctor.

It took no time at all for my teachers to decide that they would let me determine what I wanted to do, and what I wanted to be, in my own way. They had no reservations whatever, they did not care to watch the processes that would be necessary for me to reach a decision. Briefly, they did not want me around. In order to accomplish their purpose, I was unceremoniously busted, in the first semester, before I had time to get my bed warmed up.

There was no fatted calf waiting when I was returned to my home, and after a few days of an arctic atmosphere I determined to get the hell out of there and go some place where my talents would be more welcome, and appreciated. Without hesitation I decided to follow Horace Greeley's advice: "Go west, young man." And to the west I started, without announcement. The few dollars in my pocket took me by train to Albany, but that wasn't west. I was headed much farther from home than that.

Pearl diving behind an Albany lunch counter, not far from

the railroad yards, solved the food problem, and the few dollars I earned in addition to meals, were enough to buy a ticket to Syracuse. Grape picking in a nearby vineyard brought me the dollars I needed for a jump to Cleveland. I wasn't making fast progress, but I was on my way.

A convenient snow storm had put enough snow on the ground to demand shovelers to clean some of Cleveland's streets. The money I earned bought me a ticket to Chicago, where I landed at night, cold, hungry, and stone broke. From the Salvation Army soup kitchen I drifted to the Chicago Northwestern depot where the benches were hard, but the air warm, a relief from the almost zero temperature outside.

Two stiffs sitting on the bench across the aisle, after carefully and thoroughly looking me over, came and sat down, one on each side. They were friendly, but curious. From the sum of questions, answers, and conversation it developed that they were hard-rock miners and were on their way to Cripple Creek, by way of Denver.

The stiffs pumped me almost dry for my story of "how" and "why," and had gotten a good idea that I was leaving a lot of things untold. The name of the older was John T. Harrington, "John T" for short; the other, Michael Sullivan, whom John T addressed as "Sully." By the time we began calling each other by working names, John T and Sully had decided to take me in tow. Without any preliminaries, I was asked to go to Cripple Creek with them. After my unhesitating "yes," John T got up and said we had better get going. The Pinkertons (oppressive private police employed by the railroad companies) had spotted us, and if we didn't move, would pick us up.

We went out, Sully leading and John T taking up the rear. I was escorted past the stockyards, through a maze of railroad yards, and around or between box cars. We ended up in the jungles miles, it seemed, from the depot. My eyes had become geared to darkness, and I realized we were heading for one of a multitude of shacks. How Sully picked out the one he and John T were using I don't know; they were alike as peas to me.

The shacks were built of ties laid one above the other, and one could stand up inside. Cracks and corners were plastered with mud; a tie roof covered with mud and straw kept out rain water. A gunny sack over a narrow entrance, not four feet high, kept warm air in. The shack was warm, and when Sully lit a candle, I saw a pile of partly burned sticks in a hole in the middle of the floor on which sat a can half-filled with coffee. Another can was almost full of mulligan stew, a concoction of meat, potatoes, and vegetables, or whatever was easy to get, but always food for the gods.

Burlap sacks sewn together, with newspapers between, were both bed and bedding. There were only two of these, and John T went out and in a little while came back with enough gunny, string, and newspapers to make one for me. It took quite a time, John T forcing a nail through the thicknesses of gunny and paper, then passing string through two close-together holes and tying it. When we turned in, I was surprised at its warmth, and I slept like a log.

We were up before dawn. Sully took me in hand and showed me how to tie newspapers around arms, legs, and body to keep wind out and body heat in. The paper was stiff and the mess was hard to get around in at first, but it was warm and no overcoat ever gave better service. It was not long before I became used to it.

There was nothing much to pack. The gunny bedroll, "bindle" John T called it, was rolled up with our few possessions inside. A tooth brush, comb, soap, a towel so black I marveled why it did not rub dirt on instead of off, a few socks, were all that I had, and they did not add much weight to my bindle. The bindle was tied and held in shape by heavy cord, with a loop added for throwing the bindle over a shoulder for carrying.

The remaining breakfast of coffee and mulligan was put aside in the tins to wait for the next stiff to come along to use it. Then, with bindles thrown over our shoulders, we headed for the railroad yards. I was a fledgling member of the legion of bindle stiffs.

Our trip to Denver was to be on rods, the blind, or, if lucky, inside a box car or gondola. John T spotted a freight getting under way, and we climbed aboard at the end of a box car and waited between two cars until out of the yards. Then we worked our way to a gondola and spent the day under unrolled bindles to keep the cold of November from freezing us.

The Pinkertons had not spotted us when we got aboard, and the train crew kept inside the warm caboose and didn't care. Once in a while a brakie would pass from one car to another during the day, but they left us alone. Whenever the freight came to a stop, we would get off, on the off-station or water tank side, run around to straighten our stiff, cold bodies, then jump on again when the whistle blew to call the flagman back.

About dusk the freight engine pulled up beside a water tank, and we were off our gondola, heading for a camp near some leafless willows. We picked a shack, left our bindles, and went around to others that had some stiffs in them to get acquainted. After the visiting was over, John T went across the tracks to a section crew camp to rustle some food.

The next morning when a passenger train stopped at the tank for water, we made what John T called the blind, which to me was the end of the baggage car through which none of the train crew passed between station and station, and at stations, only in an emergency. I was in for an experience with smoke, cinders, dirt, and sand. It was fast travel and not as bad as by the rods on which two days of our trip was made. Every tank had shacks nearby, and we went off each day, sometimes in the late afternoon, when the passenger or freight pulled up to take on water for the engine. Late on the ninth day we were in Denver.

We rested a day at a Denver camp on Cherry Creek, and the boys planned to make a break for Cripple Creek the next morning. It was well along into December and we had to get going before snow made it hard or too rough. Three days by box cars brought us there, and the first day in town we landed jobs at the Portland mine. Both of the boys had worked in the mine the year before, when there was serious labor trouble, and the superintendent who put us on was glad to see them back.

After landing the jobs, we canvassed the town for a place to stay; we wanted to be together, but not in a bunk house. A boarding house with a large room and three cots was finally selected, and that chore was over. We were fortunate in our selection, for it was not only a clean place and the food good; but also the landlady's husband was a shift boss, and it turned out that we had been assigned to his shift.

Supper over, we went down town to the union offices and the boys paid their back dues to bring themselves up to good standing in the International Workers of the World, and in the Western Federation of Miners. Sully told me to join and I did. It was something I never regretted, and for years after I had left Cripple Creek, I paid my union dues regularly.

Thanks to the landlady's shift boss husband, John T and Sully were paired in a drift where drilling was single or double jack, and they managed to get me on as their mucker. I have never decided which was the easier, single, or double jacking. Single jacking was one-armed swinging of the short-handled, four-pound, iron hammer, while turning a drill steel with the left hand a fraction of an inch after each stroke. Two-armed swings with the long-handled, eight-pound hammer, while a partner turned the steel after each stroke, was double jacking. Either required fifty or more strokes a minute to be effective, from every position excepting standing on one's head, and in all directions—up, down, at an angle, or to one side. Single jacking was a one-man job with no resting, but the double jacking gave some rest with striking and turning being alternated at one or two minute intervals. Neither were jobs for softball players.

It was hard working by candlelight that flickered every time one moved, but there was nothing else used for light. The mine operators issued three candles to a shift, sometimes four at the better mines, often only two at the penny-pinching outfits. If the candles issued burned too fast, one had to work alongside a stiff who had a candle left, or in the dark.

It was harder getting used to the smell of dead powder smoke and the reeking, water-soaked timber, but I did. Possibly a nose broken in football rendered my sense of smell less acute. The destruction of my smelling apparatus stood me in good stead when I went in later years to live in the Far East where I was the envy of every foreigner who had a sensitive nose.

The boys had made up their minds to make me a hard-rock mining stiff, or to die trying. They weathered it, all right, but I almost fell by the wayside. It was damned rough going for a city kid who thought that, because he had played football, he was tough.

Sully was always the teacher, except in double jacking, which John T attended to. Sully began by showing me how to muck—to use the word "shovel" in a mine would be sacrilege. Muck or shovel, it was all the same to me. If sixteen one-ton mine cars were not filled during a shift, to be taken out by a trammer, it meant that a time-check would be ready in the morning and the mucker would be on his way down the hill talking to himself. Mucking put blisters on my hands until they were raw and muscle pains in every part of my body that kept me from sleeping, but under expert teaching those tortures were soon over.

After mucking came instruction in single- and double-jack drilling. Usually hard rock stiffs could work double or single as they chose. Both types of drilling were adapted to everything but narrow or low-roofed workings, where single jacking was necessary. Double jacking was impossible in narrow stopes less than sixteen to eighteen inches wide, where the holes to be drilled were mostly overhead anyway. In narrow stopes there was hardly enough room to swing even a single jack upward without wearing the skin from the striking arm by scraping it against the walls. The narrow stopes were called "Cousin Jacks," because only Cornishmen were suppossed to be able to work in them, but John T and Sully both did, and they taught me how to do it, too.

Both boys took a hand on the course in single jacking. They taught me how to point drill holes, and why; and to drill without going-by and breaking my hand, if the single jack missed the steel drillhead. Pointing each drill hole was damned important. Swinging the ordinary four-pound single jack fifty times a minute, for hours at a time was bad enough, but John T trained me on what he called a "Dago" which weighed four and one-half pounds and had a tapered face smaller than the regular single jack hammer head. The fewer holes one had to drill, the less work there was to do: drilling with the "Dago" and pointing holes so that each would break out the most rock was the answer. I didn't like work, so I paid attention to everything the boys told me, and had less work to do.

A hard-rock mining stiff had to know how to do everything that was to be done underground, so the boys gave me the works. I was taught to lay track, test the roof and walls for loose ground, and to stand clear of rock falls while I was doing it.

After I had learned the things needed to hold a job, and keep alive while doing it, there came a course in timbering, cutting hitches for stulls, framing for chutes and raises, and putting in tunnel or drift sets with room behind the posts and caps for lagging, so that weight could be taken off and rock pressures relieved. Going through one piece of bad ground gave the boys an opportunity to teach me how to spile, and put in sets under the spiling. I know of much easier and safer work. About the only thing in timbering I wasn't taught was square setting. The boys missed a bet there, and it cost all of us a lot of work, worry, and time when I decided to learn about square setting without them.

John T took me in hand on double jacking. He told every stiff in the mine that he was going to make me good enough to drill in the double-jack contests in Butte, or anywhere else that I wanted to go; and he did.

Skills in the labor part of mining were important enough, but more important was skill in loading and blasting. I learned to cut fuse, and to crimp blasting caps over the cut fuse with a knife or point of a candle holder without blowing my fingers off. Sully taught me how to make primers of these and dynamite or black powder, and to get them tamped into a hole without blowing myself up. No matter which it might be—dynamite or black powder—it was equally dangerous, and always called "powder."

Loading, tamping, and spitting was taught as an art. There was no "maybe" about it: a round must be loaded and fired right to get it to break the bottom of every hole and get a clean square face. Any mistake would mean an uneven, pocked, gun-holed face, and extra drilling to square it up again. Our rounds were always well loaded and fired; we were all too lazy to make ourselves the extra work it took to square up a face.

Sticks of powder were cut, several times, from end to end, then so tamped into a hole that the powder could be pressed against the sides of a drill hole. The tighter and harder the better. There was no danger in how hard the tamping was done before the primer was set. More care was needed with black powder, but it never had to be tamped as hard as the dynamite.

After powder was rammed home, a primer was set, and tamped firmly and tight, with care. Then more powder was rammed against the primer, and, on top of the last stick, mud. With its blasting cap and fuse, the primer was sudden death if not treated with proper respect. Sometimes, when the rock was hard, or the face tough to break clean, no mud was used and powder rammed-in up to the collar. I was taught to "load and tamp with powder" only when it had to be done, for overloaded holes loosened the roof and walls and made them dangerous.

Nothing was drilled into me so thoroughly as the necessity to use a wooden stick, never a cleaning spoon or other metal, for tamping. If metal were used, it might throw a spark and fire the hole, and for the tamper, that was it! But I noticed the boys didn't follow the advice of their own caution, for they often used a metal cleaning spoon for tamping. Nor did the fear that the boys tried to drill into me take deep root.

Almost as important as loading was the way a round should be spit, "light the fuse" in tenderfoot vernacular, and fired. I quickly learned to cut the spitting end of a fuse wide, and deep, and to shake the fuse powder into the V of the cut. The fuse spit faster when that was done right, because there was more powder to take fire, and it was seldom necessary to waste time making a new cut and try spitting again; one could get away sooner from a face that was ready to go.

Spitting with candles was something. The backfire of a spit fuse would blow out the candle, and it took time to light it again, all the while the other fuses that had been spit were still burning and getting nearer to letting all hell loose. Even snuffs, those two-inch or less pieces of burned candles that were lighted and set close by, didn't help much, for it took time to reach for one to go on spitting. The danger of heart failure ended when I learned to make notches in a short fuse and when fire came from the notch, spit with it; it was fast and sure, and I could walk away from a spit round, not run.

Several drift contracts were open about the time I was getting well along in my hard-rock education. The boys thought we could pick up a nice piece of money, and they talked it over with me. A contract was taken and we went to it. My senior class studies were under way. I was to get everything except square setting.

Regular shifts in the mine were ten hours. We had to put in more than that to make our contract worth while, and we worked out a schedule by which we put in twelve hours each, and by so doing accomplished the work of four men. The boys each took a regular ten-hour shift, and then worked an overlap of two hours at the beginning or end of the other's shift. My shift was six hours at the end of Sully's and six at the beginning of John T's. There were only two hours on either of the shifts when one man worked alone.

The ground was easy and the faces broke clean. We drilled, shot, and mucked-out two rounds a day, except when we had to timber. We were making so much money that some of the other stiffs wanted to try it too. Some of them did, but twelve-hour shifts wore them out and they lasted only a few days.

It was not only strenuous work, but we were not getting much rest, because sleeping hours were short and badly broken. John T and Sully were having the time of their lives putting me through the hurdles of my "senior year." Often I wanted to call it deep enough, but I didn't let the boys know about that. I was learning a lot and was going to stay with it if it killed me.

We finished the three-hundred-foot contract in six weeks and drew our checks. John T and Sully declared my apprenticeship completed and put on a celebration to honor the event. Wildflowers were blossoming all over the hills and trees beginning to bud or to leaf. It was springtime in the Rockies and getting along into summer in the valleys and deserts below. Each of us had a healthy grubstake, and we agreed that it was deep enough and time to move on.

Sully decided he would head for Idaho, the Boise Basin, where he would get on with one of the placer outfits. There was a lot of water, too, and Sully wanted to lie under the big trees after a shift and take it easy, when the swim he would take was over. More than that it would be cool during the summer, and Sully didn't like hot weather.

John T didn't like surface mining. Sully couldn't persuade him to go along. John T would head for Pioche, in Nevada, where he could put in the summer underground, in cool drifts. As for the nights, Pioche was high and only a few would be too warm. Each of the boys tried to get me to go with him, but I had made up my mind to go to Goldfield, in Nevada, where one of the world's great gold camps was going full blast. It was farther west, and I wouldn't be satisfied until I had covered the entire distance to the Golden Gate.

We rode in coaches to Denver, our grubstake was big enough to permit that luxury, and parted to go our separate ways. The night before taking off we staged a celebration, although there was nothing we had to celebrate. Breaking up the trio that had come to mean so much to each of us was not pleasant or easy. We probably indulged in too many "straight, whiskey chaser," for we pledged, time without number, that we would keep in touch with each other. How we could do that was not mentioned: we knew we were pledging the almost impossible. None of us expected ever to see the other again.

After the celebration, and just before we parted, Sully gave a long dissertation on jobs. There were plenty of jobs to be had, he told me, if a stiff was willing to work and earn his pay. If he didn't, he would soon be walking down the hill talking to himself. Give the boss a break, even if he was a no-good skunk. Never be afraid of putting in overtime without pay, if necessary, if it meant getting a job done. John T and Sully did it, and I tried, too.

Sully drilled it into me that I should take jobs that would teach me something, and never one I thought I wouldn't like. Most important was not to stay on a job too long, three months was about right, but never longer than four. If I didn't learn all there was to know about a job in four months, I would never learn at all. If I took his advice, Sully told me, I would learn more than ever came out of books. With that out of his system, the party was over and the trio broke up.

I tapped my grubstake for enough money to buy wool blankets and a tarp to roll them in and make a bedroll. A few good pairs of digging clothes, heavy long-handled underwear, and shoes completed my purchases, and I was ready to go. My outfit was a lot different from the one I had left Chicago with seven months before. As I thought that one over, I realized I never again would be a bindle-stiff, and something about the thought made me sad.

In resentment I determined on another fling at the blind or rods. I shipped my bedroll with my clothes in it to Reno and held out my bindle, which I had kept in case I should ever need it again. Then I went down to the railroad yards and waited for the right freight to come along. I was in Reno a week later.

It wasn't the money this time. I had plenty and could have ridden on any train I wanted. It was just youthful sentiment. I hadn't gotten used to breaking loose from John T and Sully, and I wanted companionship. One could get it from the stiffs on the blind or rods, and in the shacks near the watertanks.

The rest of the summer and into the new year I worked in the Combination, in Goldfield, and did my share of work, and highgrading, with the rest. Whenever it was possible, I paired off as one of a double-jack team, both underground and in the "open for all' contests on double-jack drilling. My partner in the drilling contests was Jack Commerford, a hard-rock stiff who knew John T and had worked with him at mines in Colorado and Utah.

We did all right in the double-jack contests. Once we were within a half-inch of winning, but we never were worse than third.

Goldfield had some of the best double-jack teams in the country. It was good practice. John T had given Jack a lot of the good pointers that he had given me, and we put them together and did all right. I was going to be in those double-jack contests in Butte, or know the reason why; and maybe I would run into John T and Sully when I was there.

The lessons John T and Sully had drilled into me in Cripple Creek had been well tamped, for I held my place with any of the hard-rock stiffs on the job. I know the shift boss was surprised that I lasted. When I struck him for a job, he was dubious as hell, but he put me on, and last I did. What influenced him was a look at my hands—they were so hard from callouses that a nail couldn't be driven into them. That did not mean that I could hold a job as a hard-rock stiff, but he knew the callouses had not been gotten from playing ping pong.

After I had put in a few shifts and knew the shift boss wasn't going to give me my time and send me down the hill talking to myself, my mind turned to that old mossback of the ivy-league college that had busted me. He would have been surprised; no, "surprised" was not it, he would have been "profoundly shocked." Nothing that I could possibly do would have surprised him.

APPOINTMENT FULFILLED

Goldfield was the last of the great gold boom camps and had about reached the pinnacle of its productive new wealth when I landed there. It did not differ much from others of the great mining camps. There was new wealth being taken from the ground and the atmosphere was tense with mining tradition. Otherwise there was no great difference between it and other frontier towns. They had something in common that was found nowhere else, and yet all had individuality shared with none.

There is little left now of the mining boom camps, or of the old frontier settlements, of their atmosphere, color, or reality. There are only legends, a few of the old buildings, possibly, and memories that still live in the minds of those who helped create the traditions.

In Goldfield were characters from all parts of the world. The hard-rock miners and the other working stiffs were the foundation and hard core. Without them the mines would have closed down, and there would have been no Goldfield. The prospectors who had located the ground under which the mines were found had long since sold their claims and moved to greener pastures.

Adventurers there were in search of a fortune, which only a few made. There were business men from the East and the Pacific Coast, wanting to take a flyer, but for the most part being taken. There were promoters whose shrewd manipulations made grubstakes for themselves, but milked dry the savings of the credulous who wanted to become wealthy overnight but lost it all. Mining engineers who came didn't believe what they saw, but couldn't do anything else because the gold was there.

Gamblers, pimps, prostitutes, and the hangers-on of sporting houses, gambling places, and saloons, none were missing. Welcome were the girls of the line, who gave solace and some brief companionship to men, the hard-rock stiffs, mule skinners, construction stiffs, and others, whose lives were toil, and often danger. There were few women other than those on the line, in the early days of a camp, and to see a woman did something wonderful to lonely men, and it was not the momentary pleasure in a bed. I was one of them, and I know.

The girls of the line were real women beneath their coldly-warm exterior, and not many were the hard-shelled harlots they have all been pictured. Many fine and substantial men married some of them, and I have not known of one such marriage that failed, or that was not a happy one.

Mostly the hard-rock stiffs were old timers, some of them Cousin Jacks—Cornishmen to the uninitiated, of whom few were better—Irish, and a scattering of others. They were loyal friends with a code all their own. If one needed help, or a friend, nothing was too much to do for him, but he had to have earned and deserved it first. Honest and trustworthy, excepting in the matter of gold, where nature had put it in the ground, everything else was inviolate.

High-grading was a highly organized occupation in the Combination and others of the high-grade mines. Far from all of the highest high-grade reached the hands of the mine operators, and many were the efforts to stop the practice. One of the efforts of the mine operators was an appeal to the clergy for help.

At a Sunday service I attended, shortly after the appeal was made, the preacher extolled to his flock the virtue of honesty and denounced the sin of stealing. The sermon was punctuated with vivid descriptions of torments in hell—fire, brimstone, and boiling oil—as penalties for breaching the Commandment, "Thou shall not steal." I had visions of high-grading becoming something of the past and being ended forever as his flock, deeply attentive, listened in astonished surprise. However there was obvious relaxation, and a sigh of unrestrained relief, when he closed the sermon by adding: "But gold belongs to him wot finds it first." I admit I was somewhat relieved myself.

In Goldfield I met unforgettable characters, Jack Commerford, Bob English, Jack Coyle, and Pete Lawrence, all hard-rock stiffs with whom I shared stope or drift faces numberless times; and often paired with one or another of them on a double jack. There were others, too, but these four and I were often to cross trails and meet again. Jimmy O'Brien, a hard-rock stiff gone wrong—turned newspaper reporter—was another. Charlie Waring was not long in graduating from the single-jack stage to shift boss and later superintendent, but often took time out from executive work to drill a hole or two, or help muck a round. The stiffs loved him. We spent December and the holidays in Los Angeles, going there by the Death Valley route.

Of mining engineers, real and self-designated, there were many, but most of them were too far up in the clouds to have truck with ordinary hard-rock stiffs. They would have done better and learned more had they come down to earth. There were a few real ones that I met, others, too, whose trails I did not cross.

The ones I did meet were great scouts. Two were C. W. Wheelock and J. Nelson Nevius, both from New York. The latter was chief geologist for the State of New York and never quite believed what he saw in Goldfield. Another was Otto Schiffner, who came down from Tonopah frequently, and in later years operated mines there. There was Otis Eubank, who often came down with Otto; Eubank had been in Greenwater and some of the other camps and got around more than Otto did. Otto finally ended up in Nevada City, California. I never learned what became of Eubank.

There was Jesse Knight, from Salt Lake, just out to look things over. Years later we were doing business with each other in Goodsprings, Nevada. Jesse had the distinction of siring Goodwin Knight, a governor of California. There were also L. C. Penhoel and H. B. Menardi, mining engineers and operators who did more than well in the mines they worked. In later years I was to be associated in business with these men also.

There was Tom Campbell, a rising young politician out of Prescott, Arizona, who was to become governor of his state not many years later. I was to be Tom's mining engineer within a few years.

The number of prospectors was legion, and among them was John Lamoigne. John drifted in and out of Goldfield as he drifted in and out of other places all over the desert. He was an old, old timer, if there ever was one. He typified them all, and added something besides. He belonged to the desert and was as much a part of it as were its sand and rocks. The desert would never again be the desert when Old John left it forever.

There was Shorty Harris, a prospector, going no place in particular, unless it was to search out someone to grubstake him for prospecting some new outcrop he had just located. Shorty was looking for another Goldfield. He thought he had found it one time, and so did a lot of others, and a boom got under way at Bullfrog.

Shorty's gold outcrop at Bullfrog gave out too soon, and with it his boom camp, but there was activity and thousands cluttered up the place, and Shorty was happy for a little while. He never stopped trying, though, and gave up only when Death Valley was movie-modernized and had lost its identity as a part of the desert in a civilization too fast for him. Shorty tried hard to become a dude-camp prospector to fit the movie picture of something he never was, and never could be. He did well enough though to satisfy those who did not know the real thing.

Shorty didn't like his movie role, died of a broken heart, and was buried in his beloved Death Valley. He knew the real desert and Death Valley were gone forever, and he couldn't take the movie version.

And not least, in color earned mostly by publicity, was Walter Scott, who was to be known to posterity as Death Valley Scotty. Scotty was a mule skinner who had freighted from his home diggings, Barstow, to Goldfield and made it his headquarters because he liked the place.

Scotty was reputed to have an inexhaustible gold mine in Death Valley. He never told anyone where it was because it wasn't there; Death Valley never could have produced as much gold as Scotty let it be believed that he brought out. If there was really any gold ore that Scotty took out of Death Valley, or somewhere else, the best bet is that he had stumbled onto some big cache of Goldfield high grade and the cachee, having died or departed for parts unknown, never got back to dig it up. It could have been that Scotty did some high grading on his own, or bought a lot of it, at bargain prices, and took the ore to Death Valley and cached it where no one would look. He did have high grade gold ore, from some place, and as late as 1916 he drifted into the Crampton and Crampton assay office in Goodsprings, Nevada, to have it worked into bars. It looked a lot like Combination and Mohawk high grade from Goldfield. The answer was "no" as Scotty knew it would be, but hoped otherwise.

Whether it was a gold mine, cache, legend, or someone who handed Scotty money for purposes known best to Scotty and himself, the real story has not yet been told. Whatever it was, Scotty made the most of it in his own peculiar way. Scotty never belonged to the group of prospectors or old timers who frequented, and lived in or near, Death Valley, as did Shorty Harris, John Thorndike, John Lamoigne, Pete Aguerreberry, and a host of others. His name, Death Valley Scotty, did, however, give him color and publicity, which brought fame to Death Valley and publicized it as one of the greatest of nature's scenic wonders. To Scotty belongs the credit for letting the world know about the place we, the prospectors and old timers, knew and loved.

When spring rolled around, I had a grubstake that was burning my pockets and crying to be spent, so I headed for San Francisco. I landed there a week before the earthquake and fire and was getting well under way when all hell broke loose. I was one of the few lucky ones, though. My quarters were a two-room apartment on Pierce Street, a few blocks west of Van Ness and out of the fire zone. I had not banked my money, and my bed roll was in the apartment, so I was free to do whatever I chose.

I had seen almost no paper money since leaving Chicago, over a year before; everything was gold or silver, with an occasional and very rare copper coin. Hard money wore pockets out, even when they were canvas lined, and the weight bouncing against legs was too much for comfort. However, I had decided to carry my money with me in a money belt. It was a lucky decision, for after the fire most of the banks were gone and those that were left didn't open their doors.

I was asleep when the earthquake hit. In a wakening daze it felt like heavy deep-mine blasts, but when the rumbling kept up I knew what it was. I dressed and was on the street in no time flat, to see what was going on. People were milling around all over the place. It was not long before smoke from fires downtown and in the Mission District reached up into the sky, and it looked bad.

With a couple of the less timid from homes near my apartment, I made it down into the city. It was a shambles. City police and troops from the Presidio were doing what they could to preserve order, stop looting, and help those who had been driven from their homes. By afternoon the fires had spread, and the worn-out police and troops had a tough job getting people out of the areas that were certain to burn.

Before dark I made my way back to my apartment and then went out again to watch the fire from open ground on a hill near Franklin and Sacramento Streets. By the time I reached the hill, hundreds of refugees were crowding the streets, some stopping at the open ground on the hill, others going on to Golden Gate Park. Late that night I took three families with children, the youngest in each family a baby, and four lost children, to my apartment.

Furniture was pushed against the walls and piled up to provide all of the sleeping room possible. There were eighteen who slept in those two small rooms that night, on the bed, under the bed, and in any place there was room enough to squeeze into. Blankets from the bed, my tarp, and bed-roll bedding were enough to keep the youngsters covered. The rest of us did not suffer; there were enough bodies in the rooms to keep us comfortably warm.

I spent the next ten days helping the Salvation Army, and taking care of the three families. The parents of the lost children had been found. Relief of all kinds was pouring into the city, and the refugees were getting food and otherwise being cared for. It was deep enough for me in San Francisco. I had done my small part. I gave the families a few dollars to help tide them over, and left them in the apartment with three months rent paid. My worries for them were over.

It had taken about eighteen months to keep my appointment with the West. I had come from coast to coast; no one could say that I had not gone West. I had not written or heard from anyone in the East since I pulled out so unceremoniously. Embargoes on train travel were off, and I decided to head for New York, pay my respects to my embittered family, and drop in to see some of the easterners I had met in Goldfield. I shipped my bedroll to Jack Commerford in Goldfield and wrote him that I would be back before the summer was over.

It was almost the middle of May when I reached New York, and again, no fatted calf awaited the prodigal's return. That I had disgraced the family through my summary dismissal from the ivy-league seat of learning remained uppermost in their thoughts.

That I had departed without notice, and had not written in all of the time I was away, was not mentioned. The proverbial doghouse had nothing on the situation I was in.

By July the family situation had smoothed out, but there were a lot of fingers crossed, including my own, and everybody waited, tensely, for something new to happen. It kept me maneuvering to hold the fort. The fact that I seemed to have plenty of money was a puzzler. The family were sure that I had turned into a bank robber and were too afraid of learning the truth to ask questions.

I told nothing of what I had done, or where I had been, excepting that I was in San Francisco during the earthquake and fire. Had I told of having traveled as a bindle-stiff, and that I was a hard-rock miner there would have been an unanimous exclamation; "Oh! A hobo and common laborer!" With emphasis on the "hobo," and "common," and the show would have been on again. Crisis after crisis had to be passed before I was ready to pull out, and the danger was over.

Two months of New York, with its canyon walls and cliff dwellings, made me restless. My grubstake, and it had been no mean one, thanks to Combination high-grade, was running low. It needed building up. New York was no place to do that. What was really more important, however, was the forthcoming fight for the world's lightweight championship between Joe Gans and Battling Nelson, to be staged in Goldfield on Labor Day. I wanted to be there to see it.

When my decision to pull out was made, it took only a few days to complete my schedule of visits to men whom I had met in Goldfield. Another three days were consumed in getting something else off my mind. What I wanted to do was to drop in, unexpected and unannounced, on that mossback of the ivy league and relieve myself of something that had been brewing for a long time. I did just that, and felt a lot better, not only for getting it out of my system, but because I found the old coot was a damned nice guy after all.

I started west, back this time, not going, to what was to be home, not as to any particular place, but just the West. Horace Greeley might not have known half of it when he said, "Go west, young man," but I at least thought I did. Traveling this time was also by train, but not as a bindle-stiff. I was a pay passenger. I missed

John T and Sully, for there was no fun traveling without them, and I wondered where they were and what they were doing.

After playing around in Reno a few days, I headed for Goldfield. Luck was with me when I stopped over in Tonopah, where some of the hard-rock stiffs I knew warned me not to go on to Goldfield. All of the tin-horns and promoters in the country were there, cluttering up the place and making a mess of it. They were getting it set for the fight fans and suckers who would come for the Gans-Nelson fight.

High-grading was getting harder to get away with, and there were more hard-rock mining stiffs than there were jobs. Prices for everything had gone sky high, as if they ever were low. The camp just wasn't a fit place for a working stiff. The girls on the line had doubled their prices, now that the free-spending blood was coming in (as if I cared).

I remained in Tonopah. It was no chore getting on at the Belmont, wages were good, but there was no high-grade to help out. The bunkhouses were fine and the boarding house chow swell. Jack Commerford brought my bedroll up from Goldfield and his own as well. Goldfield was deep enough for him until after the fight. He got on at the Belmont, and we started practicing on double-jack drilling. Before he left Goldfield, Jack entered our names, as a team, on the challenges for the double-jack contests to be held the day before the fight.

Jack and I thought we had a good chance to win, although a team was coming down from Michigan's Calumet-Hecla mine, and a couple of good pairs from Butte. Others were entered from most of the bigger mining camps, and the contests would be good. I hoped John T and Sully would roll in and get in some drilling, too. Jack thought that there would be too much celebrating before the fight, and that if we laid off the whiskey straights, we couldn't help but win.

A couple of times Jack and I took a day off and rolled into Goldfield to see what was going on. There was enough excitement in the boom camp to satisfy anybody. Hundreds had already arrived from the East, as well as from the coast, and were waiting for the fight, losing money at the gambling tables, or roulette wheels, and spending plenty over the bars or on the line. Many more came during the few days before the fight.

There were crowds everywhere, and the gangs that lined up at the saloons, gambling places, or sporting houses on the line looked no different from the ones trying to get through the subway entrances, in New York, at the rush hour. A lot of story-book wild west stunts were pulled for the entertainment of visitors. The tenderfeet were given something to remember beside the fight. Holdups, robberies, six-gun fights, men sprawled around dead or in the last agony, with ruptured bags of catsup giving the final touch of realism. It was good entertainment, and a wonderful show, but as far from anything out of real life as could possibly happen.

Jack and I took great pains to get our steel just right. We sharpened and tempered twenty steel on two and one-half inch change lengths, with ten extras on five inch changes just in case an ear broke or a bit chipped. Ordinarily fifteen changes were used, one for each of the fifteen minutes of the contest, and they were for three inch changes. Each of us had selected our double-jack hammers, long, thin and small faced. The force of each blow would be all on the steel. Handles were of long-grained, well seasoned hickory. The hammer and the handle weight combined was limited to nine pounds, by contest regulations.

Before we left for Goldfield, three days before the fight, we had gotten our strokes to sixty and sixty-one a minute; we wanted one more but couldn't make it. We were in good condition, and a lot of the hard-rock stiffs in Tonopah who had watched us practice were going down to bet on us and to watch us win. We were not so confident as they were. There were going to be a lot of good teams to go up against, but we were sure going to try to win.

Places to shack up were almost non-existent, for everything was occupied that had a roof or was reserved for incoming fight fans. Tents had been thrown up wherever there was space for them, but there still was not enough shack room, and some of the late arrivals had to go to Tonopah and come down the day of the fight.

It didn't matter to Jack or me whether we had a roof over us or not, for we had our bed rolls and we could throw them down almost anywhere since there was plenty of open space on the nearby desert. We didn't have to do that, however, for a piece of ground not far behind Ole Elliot's was the answer. It was already crowded with hard-rock stiffs, but they made enough room for two more. We threw down our bedrolls and made camp with them.

We didn't have a walk away in the double-jack contests. Mike Kinsella, and his partner, whose name escapes me, had the same idea that Jack and I did, keep sober, don't celebrate, and win. They were our only competition. It was close but we came in with a short inch to spare; we attributed it to the shorter steel lengths and bits not dulled in that extra half-inch of drilling that the three inch changes had to do.

The Tonopah stiffs who had come down to see us win had put a lot of money on us, the odds were even, and had made a killing. Jack and I put as much as we dared on ourselves, and with the winner's purse we did all right. The stiffs from Tonopah were so insistent on celebrating that we broke down and made up for lost time that night. We thought we had it coming, and it cost us nothing. The Tonopah stiffs refused to let us spend a cent; they even put bets on the wheels and made some extra dollars for us by doing so.

Mike Kinsella and his partner joined in the celebration; he was a fine stiff and a good loser. I drilled in other contests against Mike, but I was never one of a team that beat him. Mike made the championship several times and held the drilling record of fifty inches, in fifteen minutes, in the hardest rock of them all, Vermont granite. His record has never been topped and still stands.

Jack and I were lucky, for years afterwards, some stiff standing at a bar, who had watched the Goldfield contest, would point and yell to the room: "There's one of the stiffs who beat Mike Kinsella." If Jack were along it would be: "there are—," but the result was the same—free drinks the rest of the night. On what shaky foundations is fame built. Mike never used three-inch changes again: his experience with us taught him a valuable lesson.

The next day the Gans-Nelson fight came off as scheduled. No fight that I have seen since has equaled it in any way. Gans was a boxer and Nelson a bruiser. It was a fight from start to finish, and not once did either man let up trying to knock the other out. Fight fans got more than their money's worth.

Gans' boxing was perfect, and his footwork had Nelson working hard, and worried. After the tenth round Gans had the crowd with him, and booing Nelson. Nelson must have known he was licked from the beginning. He never was a clean fighter, and he poured it on that day. Sometimes Gans seemed to have Nelson where he wanted him; then would come infighting, clinches, and foul blows, Gans would slow up. The thing ended the way it should, when the referee gave Gans the decision, in the 42nd round, on a foul by Nelson. Gans wired his mother that he had "brought home the bacon," and he sure had.

The fight over, I drifted back to town and collected my bets. Odds had been good with plenty of Nelson money laying around, and I had covered as much of it as I could. It had not taken any time at all to work up a good grubstake, thanks to Mike Kinsella and Joe Gans. It was bigger than the one with which I had left Goldfield when I drifted to San Francisco just before the fire.

The next day I bought a surveying and assaying business. The chap who sold it to me had bet the wrong way, on Nelson, and after the fight had tried to replace his loss by playing roulette. He sure needed money when he ran into me, and there was no time lost in taking him up on his offer to sell.

The business was not much, for it had been neglected, but the surveying equipment was good. A shack that had a two-by-four office, a bedroom large enough for two cots, and a kitchen outfitted with pots, pans, dishes, and a stove went with the business. In one corner of the office was an assay furnace with a burner and fuel tank, hooked up and ready to go, but it had never been used. Over the door hung a sign "ASSAYING AND SURVEYING."

I ordered scales, balances, and the rest of the stuff needed for assaying. When they arrived, I was ready to do everything the sign above the door said. I had burned a lot of midnight oil since starting correspondence courses, back in Cripple Creek, but I had the unmitigated gall to start assaying and surveying without ever having run an assay or put my hands on a transit or level. I made mistakes, but there was no one to check my work but myself, which I did, and plenty of it. I put in more time getting my work right than I had in studying, at least it seemed that way. Anyway there were no complaints.

THE GIRL FROM GOLDFIELD

The assaying and surveying business prospered from the beginning. Jimmy O'Brien wrote a news item that was to have wide circulation, of a hard-rock mining stiff who had taught himself assaying and surveying and was opening an office in Goldfield. There were only a few who knew of the office I had bought. It had had so little business when the former owner operated it, that it was as if a new office were being opened.

Jimmy went so far, in his newspaper item, as to say that he had given me samples for assaying, and had had them checked in San Francisco, for accuracy. That did it. I soon had more business than I could handle.

I had no problems in assaying, the problems were concerned with gold. Dozens of hard-rock stiffs, plain stiffs, dealers from the gambling joints, bartenders, even girls from the line, brought high-grade ore, or gold concentrated from the high grade, for me to melt and refine. I refused to have anything to do with melting, and made it very clear to everyone who brought the stuff in to me, that I would do their assaying, but nothing more.

Almost every assay office in Nevada, and a lot outside, were melting Goldfield high grade, and making a nice piece of money doing it. I had done my share of high grading with the rest of the stiffs in Goldfield, but I drew the line at melting it into bars. A very fine line indeed.

One morning as I was firing the assay furnace, I looked up when the door opened. There stood C. W., a mining engineer from New York whom I had met in Goldfield the year before, and whom I had dropped in to see while I was in New York. He had a leather bag, padlocked, filled with samples he had taken from one of Shorty Harris' prospects. He was in a hurry and wanted them run right away.

The number of samples to be run that day were more than usual, and I told C. W. that if he would help me the assays would be out before dark. So we pitched in together, crushed, quartered, and bucked them down, got them into the furnace, and by mid-afternoon the weighing was finished.

There was nothing worth while in the ore. Shorty Harris had stumbled on another of his many and swiftly increasing number of poor prospects. C. W. was disgusted, for he had spent three days looking the thing over and sampling. Shorty wasn't to blame, the outcrop looked good and panned all right, but the values were not there. Shorty never had much luck with his prospects. The only really good one he ever found was at Bullfrog, but like everything else Shorty touched, it didn't pan out.

After the assays for the day were completed, and I had cleaned up, C. W. and I drifted down to the Palm Grill, put on the nose bag, and talked things over. C. W. was going to be around awhile, and I asked him to shack up with me. We went to his room, packed his bedroll and equipment, and carried it up to the office, where he settled himself.

My days were occupied with assaying and surveying. C. W. was out in the hills almost every day with some prospector who thought he had struck something good. I tended to business, and the office made more work for me than a hard-rock stiff's job, with high grade thrown in. I had picked up some helpers, both for surveying and assaying, from the drifters around town, but they were not dependable, and I had to do their work myself. I needed good help.

Someone reminded me of Charlie Taylor, a good all-around surveyor. Charlie had been in Goldfield, but had gone to San Francisco. I sent for him, and when he arrived, my troubles with surveying were over. I kept Charlie busy, in Goldfield and at other mining camps of Nevada. Art Clark, a kid who was working in an assay office in Tonopah, heard that I was looking for an assayer, and drifted in to see me. I thought that anyone who had the guts to want the responsibilities of assaying, after he had worked in an assay office as a helper, deserved a try. He couldn't be any greener than I was when I started the office. Art made good in a big way.

Many prospectors would drift into the office and ask for a grubstake. Almost always they had samples of rich ore that they told me came from some new outcrop they had found. The high-grade samples they brought in were usually from one of the Goldfield mines and easy to spot. I turned those boys down; to the others I gave a few dollars, and rarely saw one of them again.

One morning an old-time prospector, a newcomer in camp, came in with some samples that looked as if they might mean something. I ran them, and they were good enough to make me think he might have located a good prospect. The rock did not look as if it came from anywhere near Goldfield, so I gave him a grubstake.

The ore had come from somewhere around Walker Lake, so he told me, Rawhide, and the locations were on what he called Hooligan Hill. He was gone a couple of weeks and was back with some ore that I thought never came from the same outcrop that the first samples had. When C. W. looked at the samples, at first he thought he might go up and take a look at what the old buzzard had found. The ore that he had brought in came from too deep for an outcrop; to me it looked as if it were Goldfield high grade. There was nothing doing on more money for the grubstake, and C. W. was not interested.

A few days later the old timer came in again. He had been on one hell of a binge and was getting over it. He wanted to sell me his half interest in the grubstake claims so that he wouldn't have to sober up right away. I did not offer him much, but he grabbed at it, I got the deed for his half of the grubstake, and that's the last I ever saw of him. But it was not the last I heard about those claims on Hooligan Hill.

C. W. didn't have any luck although he looked at a lot of holes in the ground. He did get a lot of exercise and a fine leathery tan, but that was all. We came to know each other well and made a friendship that was to last many years, until C. W. went on the last trip of them all.

Our evenings had gotten to be a deadly monotonous routine. If we finished our work early, there were a few games of cribbage before supper. After supper we would drift down to the Palace, Mohawk, Ole Elliot's, Tex Rickard's "Northern," or one of the other places. We would play a little stud, sometimes twenty-one, or hit the faro bank, have a few drinks as the evening wore on, then go back to the office and turn in.

It was a hell of a way to waste time, but there was nothing else that we had thought of to do. In the older camps, where things had settled down, it was different, there was more visiting around and more social life. But Goldfield was infested with tin-horns and promoters, who did not believe in orderly living, and who interfered with the regular stiffs and old-timers who wanted to get things settled down and smoothed out.

One evening the routine was too much for us, something had to be done about it. The incessant rattle of ivory balls on roulette wheels, the clicking of dice on crap tables, the noises and bar-room smells. We needed something different, so we headed for the open air to talk it over.

To go back to the office was out. I worked there all day and slept there all night. We needed change. There was nothing left for us to do, but wander down to the line. There were women there, and we were tired of nothing but men. We were going to break the deadly routine or know the reason why. We should have thought of the line sooner.

We made every one of the dance halls, and parlor houses. They were about all alike, and filled with men. There was not much improvement over places uptown, excepting there were also women around.

We finally gave up and started back to the office. In getting away from the line we walked in the other direction from which we had come in, and passed through the street where cribs were located. We had not thought of them before. At least there was only one girl to a place, and no men in any of them, unless the girl was busy, and then the crib door was closed. There were a few men, wandering from one crib to another, and being orderly and quiet about even that.

As we drifted from one crib to another, the girl in the window or doorway would invite us in. When we hesitated and talked more than a minute or two, and did not go in, we were asked to move on. Our standing around interfered with what might be more profitable business.

Even with the improvement of the cribs over the other places, we were getting nothing from our evening. We had about exhausted everything on the line and were about at the last crib when we stopped to talk to the girl inside. We talked a few minutes and were asked to come in. Both of us hesitated a moment, then kept on talking. Again she asked us to come in, and a minute later told us that we must move on. But I had no intention of ending it that way.

C. W. did not want to leave any more than I did, but after all business was business—if we didn't go in we would have to leave. I asked the girl what she charged for a trick. When she answered, I asked her how much time that could take, and she told me. Without giving C. W. a chance to say anything I told her we were buying an hour; we walked in.

As soon as we had crossed the threshold she pulled down the window and drew the shade, locked the door, and collected for the hour. Then, still without a word or any hesitation, she started to take off the few clothes she had on and, when she was down to almost nothing, she broke the silence by asking that one of us go into the back room, and wait, then come back when she called.

I told her to get her clothes back on again, that all we wanted to do was talk. She looked at me, astonished, but said nothing and started to dress. Strange things happened in Goldfield, but nothing like this. C. W. did not move, nor did he say anything, but his face told me that he thought I had lost my mind.

It was the girl who first regained composure, and in a matter of moments, the line and crib were forgotten, and it was as if we were in her home, visiting. When it came time to leave, she asked us to stay a little longer, there would be nothing more to pay, and we did. When we walked out, she pulled up the shade, opened the window and waited for her next customer.

C. W. remained in Goldfield a few weeks longer, and every night we wandered down to the line and the girl's crib. We never spent less than an hour and sometimes much more, and never left without giving her something extra. Our evenings had become fruitful and pleasant, and much more than a happy "by-chance." We didn't miss the places up-town at all.

Work piled up after C. W. left, and it was several nights after he pulled out that I went to the crib again to see the girl, alone. The girl was glad to see me and asked about C. W., and why he had not come along. After I told her that he had gone back to New York, we talked of other things.

It was much nicer going down to see the girl without C. W.

along. I was a regular visitor. Even extra work in the office did not break the new routine. I went down at the same time every evening, early, so as to reach the line before many of the stiffs from up-town would be there. A lot more time was spent with the girl than in the days when C. W. and I went together. I never left without more than doubling what was supposed to be paid and, although she was always unwilling to accept more, she did. Months later I was glad that I had done it that way.

Although a real friendship grew between us, try as hard as I could she would not talk about herself, or what had brought her to Goldfield. Nor could I get her to tell me her first name (no one ever asked about a last name) but she wanted to be completely without any name—even an assumed name to use when I talked with her.

One evening when I went down, she wasn't there. The girl who had taken over the crib only knew that she had left that morning. None of the other girls knew anything either, although I suspected that one or two had some idea of where she might be headed; but whatever they knew couldn't be blasted out of them, so I gave up. There was nothing I could do about it, the girl was gone. I did not make any more visits to the line after that. There was something missing from Goldfield, and from then on I spent my spare time in the office.

Within a few days there was new excitement, a discovery of real high grade, at Rawhide. I had a flock of claims there, thanks to my grubstaking that prospector. I sent Charlie Taylor up to look the ground over and to survey the claims. Things had not gotten started yet, but it looked as if a boom might be in the making. I wanted to have everything set and ready when, or if, it came.

Charlie was back in ten days; he had made the survey, filed and recorded amended notices, and then looked the ground over to plot the vein, or veins, on his map. He had intended putting a few shots in the outcrop where the old prospector had made his discovery, then decided he would leave it alone. The old buzzard, so Charlie thought, might have salted the thing. There were fragments, small ones, of high-grade gold ore lying around, that might or might not have come from the outcrop.

Charlie did not shoot the thing: if there was high-grade in the vein it would not get away, and if none was there a few shots would destroy the prospector's artistic job. If it was salted, Charlie said, it was the best job he had ever seen. Charlie didn't know and couldn't tell, but he was willing to bet a year's pay that if it were not salted, the ore wouldn't go down very deep.

About that time George Graham Rice and Nat Goodwin, a famed comedian, became interested, if not active, in stock promotions formed on Rawhide properties. They had been in and out of Goldfield and the nearby camps where someone had tried to start a boom. Goldfield high-grade had been used as a come-on, and not a few outcrops had been salted with it. Neither had run into anything worth while although they did have a few promotions that were not doing very well.

One morning as I was firing the assay furnace, Nat Goodwin and some of his Los Angeles pals came into the office with some samples for assay. The ore looked familiar, but I said nothing about that. After a lot of stalling and beating around the bush, Goodwin asked if I would sell him my claims in Rawhide. He had looked them over, he said, and thought they looked good enough to take a flyer on. I knew then where the ore he wanted me to assay came from.

I told Goodwin that my business kept me tied down in Goldfield, that I was too busy to go to Rawhide with him and close a deal. I told him that I had not seen the claims, that I had gotten them through a grubstake, and I didn't think they were a good prospect anyway. I didn't have any time to talk to him about the claims, I wasn't interested in selling, and I was too busy to talk about it until after my work was finished.

Goodwin paid no attention to what I said and kept on talking. I knew that he wanted those claims and would buy them at any reasonable price, or even one not so reasonable. Charlie Taylor took over then, and I went back to the assaying. The maps that Charlie had made were brought out and looked over. Charlie took over from there and told Goodwin that I wanted to see the ground before I sold the claims to anyone. He said that he had been up and looked the ground over and thought the ore might have been salted, and that I didn't want any part in the sale of a salted mine.

I was listening, as best I could, with the roar of the assay furnace trying to drown out every other sound. Charlie raised his voice, loud enough for me to hear, and Goodwin followed suit. Goodwin was not impressed with what Charlie had told him. The idea of anyone not wanting to sell a mine until after it had been looked over to see if it had been salted or not, didn't go over; nobody was as big a fool as that.

What had started with an honest opinion of the Rawhide prospect made Goodwin all the more anxious to buy the claims. From the way he talked I knew he had looked them over from end to end, and that whether salted or not, he wanted them. Finally he stopped talking to Charlie and yelled over to me to come over. I knew what was coming, so I fooled around on some samples to assay, for a few minutes, and went over to the counter to talk to him.

Goodwin made me an offer, and I doubled the amount and added a few terms of my own. Then I told him that that was it, I was too busy with my work to talk any more then. If he wanted to buy, he could come back after the assaying was finished, and I set five o'clock as the time I would be free. I made it clear that I had no intention of haggling over price and terms; if he didn't like them, he need not return.

For some reason Goodwin wanted those claims, and he wanted them in a hurry. Charlie called to me about four-thirty and told me to look out the window. There was Goodwin with his pals, waiting. They were impatient. Every few minutes they looked at their watches. They were going to return, on time, five o'clock, on the dot. They were within a few feet of the door and couldn't miss.

At five o'clock Goodwin came through the door, accepted my offer, and within two hours the deal was closed. I got a cash grubstake and an interest, in stock, in the company he was to form. I couldn't lose.

The prospect never panned out. The high grade was either a pocket or the thing had been salted. Either way, Charlie had guessed right. There was a short lived boom at Rawhide, but like most of the others, it petered out and from the thousands who drifted to it, only a few were left not long after.

I met Goodwin a number of times later. We never discussed the Rawhide property. There was nothing to discuss. Goodwin was having wife trouble about that time: he was shedding number four or five, to take on number five or six. His affairs were so involved, and so frequent, it was hard to keep up with them. Whatever his mining or marital affairs, he was a great comedian.

The assaying-surveying business continued to prosper, and I

had no time for much besides work. November was under way when black pneumonia struck in Goldfield and Tonopah. It was probably a virulent form of influenza, but whatever it was it was a humdinger and a lot of hard-rock stiffs, and stiffs, went over the hill to whatever reward they had coming to them.

One morning when the boys came to the office to go to work they found me with a high fever and delirious: black pneumonia. Charlie and Art took me to Oakland since the Sacramento and Reno hospitals were crowded, and I was nursed back to life again. It was a close call. When I was discharged from the hospital, I took a room at a hotel. I had written home while I was in the hospital, and it was not long before there were visitors, mostly relatives who had migrated to the west years before, and their friends. I had no complaint about not having any company.

It took a while to get my strength back, and during that time I was asked to luncheons, teas, and an occasional dinner. I didn't refuse the invitations for they helped to pass the time. The most frequent invitations came from some sort of a distant relative who insisted that I call her "Aunt Matilda." Aunt Matilda was from the New England branch of the family that came over to America early in the sixteen hundreds and just oozed tradition and formality.

Aunt Matilda's name alone warned one what to expect. I made the best of it, and wore afternoon clothes for her teas and swallowtails for dinners. Had any of my hard-rock mining stiff friends seen me, they would have thrown me into the nearest waste chute, to be poured over the dump. When it came time to pull out of Oakland, I sold the clothes to a shop that specialized in that sort of junk.

Aunt Matilda frequently had her teas or dinners at some swank hotel. Then, too, she had her "at home" affairs. Whichever it was, it was something that everyone must attend, with Aunt Matilda the hostess, provided, of course, that they were invited. Her affairs were one of the social "musts." No one dared crash one of her "things." They were so stodgy, staid, and formal that I wondered why no one did something so as to be crossed off her list. Almost every face I saw at one party appeared at the next one she gave. Almost!

At one of Aunt Matilda's "at home" teas, after everything was under control and the tension that her affairs demanded had eased, some late arrivals appeared. I did not pay attention at the moment. There were always people coming and going at Aunt Matilda's affairs, and after having been to four or five I was sure that I had met everybody on her list. I was very much mistaken.

My heart stopped beating when Aunt Matilda introduced me to an attractive and beautifully gowned newcomer—the girl from the Goldfield crib, "Mrs.—." The face of the girl went the color of wet ashes. It took both of us moments to recover our breaths and composure, to say nothing of normal color, and to come up with the stereotyped "so glad to meet you" stuff. That over, Aunt Matilda herded me over to the introduction of another of the late comers; she had been too busy being her correct New England hostess self to have noticed anything.

I am sure the girl did not enjoy that tea and probably wondered about many things. As for me, there was no enjoyment, and I left early, but not until the girl had gone, in all likelihood much earlier. I had not seen her go. I might have followed her if I had, and I was glad that she had been wise enough not to give me that opportunity.

Aunt Matilda was giving a New Year's Eve dance at the Claremont. Would I come? RSVP. I would. The evening of the dance arrived, and among the guests again, was the girl from Goldfield. This time Aunt Matilda introduced me to her husband. The girl and I had been introduced at the tea, and I was glad that it had been this way. There was no outward evidence of stupefaction or tension this time.

It was very evident that Aunt Matilda had a high regard for the husband, for she remained with our group a few minutes. Long enough for me to ask the girl for dances, and on her card I put my name down for two waltzes, one of them the midnight dance. It was an unfair advantage: she could not refuse had she wanted to, nor could her husband protest ;Aunt Matilda stood there, beaming at her "nephew," and to say "no" was impossible. I did want to talk to the girl, and the only opportunity I might have would be during a dance.

The girl danced a few times with her husband and I watched. He was not the kind of a man that men liked, but some women would. There was nothing about him that I could admire; even his dancing was poor. I did not like him.

The turn for our first dance came and we waltzed onto the floor.

She was a better dancer than I had thought from seeing her with others, light and soft and graceful. We did not speak to each other until we left the floor after the dance was over, then we formally thanked each other, and parted when I took her to where her husband waited.

Then it came time for the old and new year dance. It was the same as the one before, excepting that I held her tighter, and she rested her arm on my shoulder a little heavier. Our hands held each other's more firmly and once in a while there was pressure from one to the other. Nothing more. There was no way to say what I had wanted to say, or what I had in my mind to ask. As we danced the old year ended, and the new year came; the past was gone.

As the waltz ended, her arm slipped down my back and she held me tight, a moment, as I did her, and our hands pressed each other's hard. Then she put her hand on my arm as I escorted her from the floor, thanked her, formally, for the dance, and left her with her husband.

After the dance was over, as soon as I could, I sought out Aunt Matilda and carefully guided the conversation to the girl. I spoke of how charming she was, of how beautifully she danced, and was getting deeper into it when Aunt Matilda interrupted me. "Frank," she said, "don't talk about that woman, that—" Aunt Matilda sputtered and didn't finish what she intended to say, then added, "Don't ever mention her again."

Before I could ask why, or adjust my mind to the idea that Aunt Matilda knew all about the Goldfield period, she went on; "Do you know what she did?" Fortunately there was no pause for me to try to answer; anything I said in reply, or nothing, would have been fatal.

Aunt Matilda saved me, as she answered the question she had asked of me: "When her husband was accused of embezzlement, and was in disgrace, and about to be arrested, she left him! And she left her three children, too! Then she came back the minute she found that her husband's accounts were all right, and that he had not stolen the money after all." Then, as if I had not been sufficiently impressed, she added; "Never mention her name again! Never! I invite her only because her husband is so fine and wonderful." How little Aunt Matilda knew!

Jack Commerford, Goldfield, Nevada, fall, 1906. Crampton Collection.

Double-jack drilling contest for championship of the world, September 2, 1906. Each team drilled for fifteen minutes in Vermont red granite, alternately drilling and turning with each change of steel. Jack Commerford and Frank Crampton won with 43$\frac{1}{4}$ inches over Mike Kinsella and his partner, who drilled 43$\frac{1}{16}$ inches. In this picture Crampton is on hammer, and Commerford is turning. On Labor Day, 1908, Kinsella and his partner won at Butte, Montana, from Commerford and Crampton, 44$\frac{7}{8}$ inches to 44$\frac{5}{8}$ inches, drilling in Vermont red granite, and took the championship. Kinsella and his partner were the best of them all some years later, drilling 51$\frac{1}{8}$ inches in Vermont black granite for a world's record that has never been beaten. Crampton Collection.

Ancient hand-operated fire-engine water-pumper, sunk into the ground to its body bed and axles from long standing. Hamilton, Nevada, fall of 1906. Charlie Taylor stands beside the old fire-fighting apparatus. Crampton Collection.

The Indian shack in which Locke and the author were sheltered after they had found quicksilver and become lost. North of Jolon, California, fall of 1907. Photo by Locke. An Indian and unidentified woman stand beside the horses. Crampton Collection.

The Harris post office and store, Pleyto, California. Geo. W. Harris, left; the author, seated; E. G. Locke, third from right, beside tree; Mr. Cruikshank, second from right. Crampton Collection.

At quicksilver mine, north of Jolon, California. E. G. Locke, left, and Tom Stiles. Crampton Collection.

Midas, California, July 11, 1907. Derailment of the Overland Limited. The author was riding the "blind" on the baggage car, center background. Crampton Collection.

GOLDFIELD LOSES A HARD-ROCK STIFF

The Christmas and New Year holidays were over. I was recovered from my bout with the black pneumonia and fit to return to work once more. I had been brought to Oakland early in November, and two months of inactivity, unless Aunt Matilda's affairs could be called "activity," were making me restless. I didn't want to return to Goldfield. There was nothing there that urged me to return. My experience with the girl of the crib would bring back memories that I preferred to forget. Goldfield would remind me of her. There was nothing but the office, and I couldn't pull out and leave it hanging in mid-air.

Luck came to my rescue. In drifting around San Francisco, a few days before the day I planned to leave for Goldfield, I ran into Ernest G. Locke, a mining engineer who had drifted into my Goldfield office with samples to be assayed. Locke was a graduate of Oxford and had gone to Heidelberg for his mining degree. He had mined in Asia and South America, and had been in South Africa with Cecil Rhodes and John Hayes Hammond during the Transvaal difficulties that resulted in the Boer War. He was the only Englishman I ever met who considered that the Boers had been given an unfair break. Locke was looking for a mine for London backers and wanted me to be his assistant for a few months while he was looking for it.

My problem was solved—I would not have to go back to Goldfield. I wrote Art Clark and Charlie Taylor to come down to talk things over. When they arrived, I told them that I wanted them to take the office over and run the business for me. Art was to be in charge of the assaying and Charlie the surveying. Art liked my idea, and he wanted to buy a share in the assaying business. I took him up on that one in a hurry. Art was a responsible kid, and he would keep the office pure and not use it for melting high-grade.

Locke's backers wanted a quicksilver mine, as good as the New Almaden. Why they wanted to mine a metal as touchy and as uncertain as quicksilver Locke could not fathom, but, being a good soldier, he asked no questions and was out to look for it for them. There were a lot of quicksilver prospects in the Coast Range, north from San Luis Obispo to Monterey, and Locke planned to take a look at most of them. If he looked at them all, it would take a year. He wasn't going to do that.

It was well along in January when we got started, and the weather was rotten. We would not be troubled by the heat that plagued the Coast Range country in the summer, but we could get awfully wet from the rains, or an occasional snow.

We made camp in an unoccupied portion of a shed a hundred feet to the rear of the post office-store at Pleyto, a station on the stage line from Bradley. George Washington Harris, the owner of the outfit, who tripled as postmaster, storekeeper, and cattle rancher, helped us to get settled. Harris had a few quicksilver prospects of his own, mostly secured by him for the payment of some bill or as the result of a grubstake. There was no doubt, in our minds, that he had every intention of unloading one of them on us, if he could.

On saddle animals or by buckboard we covered miles of territory, accompanied by Harris, who was more than anxious to show us everything he had seen, or had heard about. There was nothing in the hills from south of Adeladia to the San Antonio Mission that looked good enough for Locke to be interested in. The only one we looked at that had any resemblance to a mine was the Klau, west of Paso Robles, but it was in operation and the owners would not consider a sale. They were doing well enough and thought they could make more by holding on and operating it themselves.

After six weeks of looking around in the more settled area of the Coast Range, Locke decided we would do some prospecting on our own, north and west of the San Antonio Mission, in the Big Sur section of the range. Harris let us have a couple of mules to pack our bedrolls and supplies, and two range horses to ride.

It took a week to make the upper reaches and headwaters of the San Antonio River and three more to get into the rugged country beyond. We prospected every stream, wash, draw, and arroyo, but no quicksilver, not even a color.

The only things we found were hot springs, and we bathed in the soft water flowing from them to clean off the grime. The weather had been rotten with lots of rain and for short periods the arroyos and gulches ran deep enough to make us follow the banks instead of the bottoms. There was all the water we wanted for prospect panning. There had been some snow, too, but it melted fast. Before the trip was over the weather warmed up and clothes lost their cold dampness; soon the ground and streams dried up.

Locke was getting discouraged, but wouldn't give up, and every day he would insist on going on for a few days more before going back. It was all right with me. I was adding to the theory I had gotten from my correspondence courses and getting practical lessons in geology. Locke was a good teacher, and patient in explaining intricate geologic phenomena.

After a month of day-to-day going Locke finally selected a range of hills to the west and slightly north as the last point beyond which we wouldn't go. It looked as if the crest of the ridge might be the divide between the eastern slope of the Coast Range and the Pacific. After the ridge was made, if we hadn't found anything meanwhile, we were going to start back for Pleyto, along a southeasterly route so that we would cover ground we had not passed over in coming up.

The country ahead looked as if it could be navigated by no mining stiff. Something better fitted for the job would be a cross between a hard and experienced old-time explorer and a mountain sheep. What lay ahead didn't deter us in the least. Locke wanted to find that brother of the New Almaden—a sister to it would do—and that was that.

The going was tough, even for the mules and saddle horses, and for us on foot. We had dismounted, for had we stuck to our saddles, we would have been scraped off by the branch of some tree or overgrown brush. Moreover we had to cut a trail most of the way, and to get on and off of our mounts every few minutes, to hack off some interfering branches, would have been a nuisance. It was better on foot anyway, for we watched the ground and did some prospect panning from time to time.

As a prospecting venture the trip to the ridge was not fruitful, but as a mountain climbing feat, with four-footed animals, it was a grand success. Both of us had gold medals due to us for having negotiated a distance of less than five miles in three days. We never got the medals. When the crest was finally reached, another ridge appeared beyond, another five or six miles. Despite the disappointment, it was a wonderful view, and the sunset beautiful.

We made camp under a long overhanging cliff, fed the animals the last of the oats we had brought along, had supper, and turned in. Next morning the mules and horses were gone. They had taken a dawn's-early-light look at the country beyond, had known there were no oats left, had thought we were going to the next ridge, had decided it was deep enough, and had headed for home.

The country to the west looked like some supreme punishment invented in hell for the torment of the damned, but the country we had come through from the east after we had gotten a good view of it from above didn't look much better. I didn't blame the mules and horses for pulling out.

We knew that we were in for trouble—plenty of it. We did not know where we were excepting in a general way. We couldn't follow the narrow valley we had recently crossed. The stream bed in its bottom was heading south and west; it could end in the Pacific or in a branch of the Nacimiento. We didn't dare take a chance and must go back along the trail we had blazed while coming up with the animals, before we dared follow along the bottom of any valley, toward the south.

We had planned to get back to Pleyto by giving the animals their heads and following them out. Now we were on our own and didn't like it. To pack bedrolls, or anything other than food, and not much of that, either, was impossible; so we each took one blanket and as much food as we could carry, and put the rest of our junk under the shelter of the cliff and were ready to move.

In our worries about losing the animals, and in the haste to get going, we had paid no attention to anything around us. Locke was the first to spot it, ribbons of deep-red cinnabar, running through the face of the cliff in all directions, some as wide as two and three inches. Immediately we dropped everything and prospected the outcrop for hundreds of feet in each direction. We had stumbled onto a beauty.

We built monuments, blazed trees, and posted location notices on everything that had a remote possibility of having quicksilver ore on it. After the locations were established, we put in the rest of that day and all the next in moving enough rock to satisfy mineral location requirements. The few pounds of powder we had brought with us was used up, and the rest of the work was done with picks and gad. Although there was no probability of anyone being fool enough to come into that part of the Coast Range, someone might, and nothing was omitted in making the locations valid in every way. What we had found was much too good to lose.

It was dark by the time we had finished the location work, and again we camped under the cliff. Our respect for that shelter had increased a thousand fold, not because it was as good a place as we could have found, three evenings before, but because of its mineral content. In fact we forgot all about its being a shelter.

In the morning we climbed trees and fastened our tarps as near the tops of them as we could, cutting a lot of branches away so the canvasses would not be torn when they whipped under the wind, and could be seen from a distance. It would be a hell of a place to find again, unless there was something to guide us to the spot. The tarps in the tree tops would do it. Locke wasn't going to miss finding this brother, big or little, or sister, of the New Almaden when he returned.

Four days later, after wandering down gulch or stream beds, over ridges and down hills, we were lost. No matter how hard we tried we couldn't orient ourselves. Food had not become a problem. We had bacon and a few other things with us, left from our camping supplies, and because both of us were good shots with six-guns and rabbits were plentiful, we were getting along all right.

Lack of water was giving us trouble. Although there had been a lot of rain, it had run off and not much went into the ground. Even the rock-cleared depressions in stream bottoms held only a few spoonfuls, or the wind and sun had dried them out. Our canteens had been emptied, and the few drops we found in rock crevices were not enough. About the time we really began to worry, we dropped down into a narrow valley and saw a water hole, with cattle in and around it. The water reeked, but it was water, and where there were cattle, men could not be too far away. Our difficulty, however, was going to be in finding them.

It was almost dark when we smelled smoke, and we followed the fragrance up a draw to something in which men were living. It turned out to be the home of some Indians, four adults and so many children all milling around in such confusion that they couldn't be counted. No one was glad to see us, but when we explained that we were lost, and they found out we were not game wardens, we were as welcome as we had been unwelcome before. The reason for the first sorry welcome we discovered when we turned in to sleep that night.

We were invited in, fed some tasteless food and put up for the night. Put up is right, on the log pole rafters of the ceiling of the room below. Above the ceiling was the low sloping roof through the cracks of which the stars shone and the wind blew. Over the rafters were scores of ripe and fragrant deer pelts, thrown there no doubt to keep rain water from the family below. The Indians slept on the dirt floor, and whenever we moved to try to find a comfortable position, which was so often that there was no sleep between turns, bark from the poles, and trash, would fall between the cracks and drop on the sleepers below.

The next morning we were up early and the Indians, all of them, led us to an almost trackless and unused road about two miles distant. As we walked away the Indians watched until we were out of sight to be sure they would not have us on their hands again, to drop stuff on them all night. They had nothing to fear; only the sheerest of desperation would have prompted us to return. Sleeping on those hides had given us a scent that, had a search party been out to locate us, the most inexperienced and untrained hound dog in the country could have picked up.

After ten miles of walking we reached Jolon, where we rented a buckboard, with a driver, and were taken to Pleyto. The mules and horses showed up the next morning just as we were preparing to pull out for the county seat to record our location notices. We bathed in several changes of almost boiling water and burned our clothes at a good distance from our camp. By midafternoon we had recorded the location notices at the county seat, Salinas, and that night we were safely in bed in a San Francisco hotel.

I did not see much of Locke for a week. He was busy sending and receiving cables and writing letters. One letter, mailed to Harris at Pleyto, asked him to put on enough men to have a wagon trail road made to the quicksilver locations. Locke didn't want to wander all over the hills and canyons in order to reach them when he returned. Instructions finally came from London for Locke to develop the prospect, and we got busy purchasing equipment and the things required in starting work. Locke went down once to look things over and was back in a few days with our bedrolls, the tarps none the worse for wear as signal flags.

While Locke was at the quicksilver prospect, I ran into Jack Commerford. Jack had left Goldfield and had drifted to Grass Valley where he worked for a time in the North Star. He told me of a couple of Cousin Jacks who thought that they were good, and boasted that they were going to Butte to drill in the contests to be held on Labor Day. They were second-raters, according to Jack. Would I team up with him, go to Grass Valley, get a job, and work up an argument with them so that they would challenge us? Jack's idea sounded good. There was a chance to pick up a few extra dollars and have fun doing it. I was soft, though, and needed work to harden me up. In addition to that we needed practice before we could take them on with any chance of winning.

When I told Locke about Jack's idea, he got a great kick out of it. He had watched Jack and me win the contest in Goldfield the day before the Gans-Nelson fight, and there was no doubt in his mind that we would win. Both of us went down to the quicksilver prospect, Jack as a shift boss and myself as an assistant superintendent. Jack and I got the hardening we needed, but we had no way of telling how good we might be. The rock that we did our practice drilling in was so soft that, had it been hard granite, all world records for drilling, even with machine drills, would have been broken.

By the time the middle of June rolled around we were hard and toughened up again, so we pulled out with the blessing of Locke and the hard-rock stiffs on the job. Locke promised us that he would come up, and if any of the stiffs from the mine wanted to go along he would pay half of the railroad fare. He told us that, if he was any judge of Cousin Jacks, we would need someone around to stand up for us when we won. Every last stiff on the job, including the cooks, took Locke up on his offer.

Jack and I went to Grass Valley and got on at the North Star. We said nothing about what had brought us there and let nature take care of itself. It did. Those Cousin Jacks were a clannish lot, but they broke down when they got to talking about their "champion" double-jack team, and of how good they were. From the manner in which the virtue of the "champions" was extolled, nothing to equal their prowess had ever appeared before. In reality the boastful "champions" had never been in any contest to prove that they were the best double-jack team in Grass Valley, but their followers had been so enthusiastic that the other teams didn't dispute the claims.

Jack got his pick in and was swinging it hard. He had never heard of any double-jack team that good, and he had known them all. He thought that if he could get a partner, he could out drill the champs himself, and he would do all of the drilling and let his partner turn. His remarks sounded as if he were boasting, and his banter when one of the Cousin Jacks tried to pin him down, and he refused to talk about it any more, had them sure that he was a fourflusher.

It ended in Jack's accepting a challenge for a double-jack contest, "if" he could find the right partner. Well, that was easy, the Cousin Jacks would find one for him. There was nothing doing on that, Jack told them that he wanted someone he knew. Did he know anbody in Grass Valley, or the mine? Jack did, and pointed to me. He would take me if he could get no one else. Meanwhile he would look around to see if he could locate somebody.

Jack's looking around gave us the opportunity we needed to go out and practice in the granite that we would have to drill in. We had our contest steel all sharpened. It had been brought up from the quicksilver where we had sharpened and tempered it for the drilling contest that we hoped we would get. We had a lot of extra steel, too, that Locke told us to take along, to do our practice drilling with. He and the stiffs with him would take it back when they returned to the quicksilver after the contest.

Drilling was to take place on a Sunday, and the preceding Friday Jack let it be known that he was going to have me drill with him. He could find no one else. Either way the Cousin Jacks looked at it, they couldn't lose. Should Jack do all of the drilling, he would tire; two men drilling against him, on changes, could not help but win. Should both of us drill, on changes, it would be still easier.

The backers of the two boastful stiffs were so sure that their champions were going to win that they threw a celebration for their expected victory that night. Jack and I went from place to place, wherever a good crowd was celebrating. We got a good sized side-bet, at odds that would make a stake for us if we won. The Cousin Jacks took a good look at me, a bare twenty-year older, and not much more than "ten stone" weight, and decided their team couldn't lose. We also made bets over the bars at even better odds than our side bet.

We were not taking advantage of the Cousin Jacks. We had used our right names, and if those hard-rock stiffs did not remember that we had won the contest at Goldfield the day before Labor Day a year ago, that was their tough luck. Some of them certainly had been in Goldfield that day, and the day after, and if they had forgotten who the winners had been it was no skin off our noses.

At the quicksilver we had sharpened and tempered our steel for the granite around Grass Valley. Jack had worked in some of the mines there, some in serpentine, others in granite, and knew that there was an occasional crack, or soft gouge, and that there was danger that the cracks would wedge a steel for a moment and raise hell, or that a soft seam of gouge might be cut and more hell would follow. Our steel were sharpened and the ears shaped to take care of either, if we ran into anything like that. Although we expected to win, we hoped that the Cousin Jacks would not think of the same thing.

The contest was to be held between Grass Valley and Colfax. The granite there was harder. Preparation for the contest couldn't have been improved. Instead of a block of granite surrounded by a platform, on which the drillers stood for all to see, the drilling was to take place at the foot of a slope, on which the spectators could sit and look at the contestants below..

The surface of the granite in which we were to drill had been taken off until the partly weathered surface was all removed, and the face leveled off by some expert stonecutter. The only bad feature was that we would not have hoses from which water could be constantly fed to the holes for cleaning out the mud from the drill cuttings. We would have to clean our holes with water from tin cans. It wouldn't make any difference in the number of strokes we would use, but the depth of the holes would be less. The disadvantage was equal on either side, so it didn't matter.

The contest was a gala event. Everyone in the country was present to see impudent "foreigners" beaten. Locke and the stiffs from the quicksilver were the only ones who were making bets on us to win. Even so the odds against us increased, and we covered them all until we had no more money to cover any more. If we lost, the stiffs who had come up with Locke from the quicksilver would have to walk home.

The confidence of the Cousin Jacks increased until the time for the contest drew near. How could a Cousin Jack team lose against an Irishman, and a kid still wet behind the ears? The confidence was shaken when we brought our steel from a buckboard nearby and laid it beside the spot where we were to drill our hole. The champions came over to take a look, and what they saw dulled their former ardor for the contest. No novices ever came into a contest with steel with bits such as we had. Then they looked at me, getting the cans of water set in the positions we wanted them to be for quick handling, and their confidence returned.

Before Jack and I were to drill three other of the Grass Valley double-jack teams were to have a contest of their own. The "champions" and our team would be a separate event and take place last. Jack and I sat beside Locke, and as we watched the drilling, we were not so sure that we had a ghost of a chance to win. Those teams were good, but it was apparent that none had done any special practicing for the event, and all could have done better if they had. The winners put their hole down a fraction over forty-three inches, which was good drilling, practice or not. Then it was Jack's and my turn.

The starting six-gun was fired for our contest, and we were off. Jack was striking while I turned the starter. When each steel change was down, we changed turning and striking positions. We were encouraged and cheered on by the quicksilver stiffs, scattered among the spectators on the hillside. Every minute we would hear Locke's voice boom out, sixty, sixty-one, sixty, sixty-one. He was calling the number of strokes that we made each minute, and once in awhile there would be another voice calling a number, fifty-eight, fifty-nine, fifty-nine, sixty, but never sixty-one, and only a few of the sixty. It was the number of strokes the champions were getting in. If we were striking anywhere near an even weight, we should win by one half inch, at least.

When the six-gun ended the contest, we were on our fifteenth steel change. We had not chipped a steel, nor were any badly dulled. The champions had three steel with chipped ears and the steel-chips had to be drilled out with the hole. Not a crack or gouge had been hit by either team. The chipped steel-ears were tough luck for our opponents. They should have sharpened their steel better, or had a better temper. Whichever it was, it helped cut them down to size, but those extra strokes would have won for us anyway.

We had put our hole down forty-three and seven-eighths inches, the champions forty-one and five-eighths. To be beaten by over two inches was too much for those Cousin Jacks, and by an Irishman and a wet-eared kid. Those holes were measured, not once but several times, before the champions were satisfied they were beaten. Once I thought they were going to claim our hole as theirs.

When everything was settled, they tried to be good sports about it, but it had taken almost as much drilling as in the contest to get it into their heads that they had lost. It was a terrible blow to them, but the worst blow was when we went around and collected our bets. The quicksilver stiffs collected, too, and Locke was as pleased as he would have been had he drilled in the contest and been one of the winning team.

Jack and I had a roll between us that would choke a horse. It would take those Cousin Jacks, and their friends, months of work in the mines to get back what they had lost, even by combining pay with what they would pocket in high-grade. Our popularity had never been great, but it reached its lowest tide in Grass Valley that night. When morning came, we called it deep enough and drew our time, at the same time announcing that we were going to Butte for the Labor Day contests. We learned later that the Grass Valley "champions" decided not to go to Butte that year.

We had no idea that we would be good enough to win in the Butte contests, but we thought we were good enough to make a try. Locke had been so pleased that we had won that he gave us the borrowed drill steel that we had used in winning, and we took that along to Butte with us. We had done all right in the Grass Valley contest, but it was not good enough for Butte. We would have to improve a lot to be in the running.

Immediately after the contest in Grass Valley I wired Art and Charlie to meet me in Reno to discuss problems that were developing in the assaying-surveying business. Jack and I took a train out of Colfax, and when we reached Reno, I let him take my things on to Butte, where I would meet him in a few days. I kept my bindle, for enroute to Butte it might be quicker to ride the blind or rods than wait for trains.

Art and Charlie were waiting when I left the train in Reno, and we got down to business without delay. They had a telegram from Locke with them. He had sent it to Goldfield, believing that I would go there. Locke wanted me to meet him in San Francisco as soon as I had finished whatever I was doing in Goldfield. He did not say what he wanted to see me about, but he would wait until I arrived.

Goldfield was slipping from the diaper stage of a boom camp and was growing into manhood fast, thanks to the San Francisco earthquake. The tin-horns, promoters, and hangers on who had infested the place and had kept it from getting down to earth were having a hard time; in fact, most of them had deserted the place. The earthquake had tightened up loose money. Money for promotions, or other questionable endeavors, was hard to get. The transition had done Goldfield a lot of good, but it had cut the surveying and assaying business to almost nothing.

After we had talked the situation over, we decided to open an office in Santa Monica. Art was going to remain in Goldfield; Charlie had taught him surveying and mapping, and he could handle anything that came along. Charlie, already a top-notch surveyor, had learned assaying, and would have no difficulties in taking charge of the Santa Monica office, so there were no hurdles in that direction.

Art was to keep me informed as to how the Goldfield office was making out, and, as soon as he thought there was no more hope of making it pay, we would decide what move to make next. Meanwhile I was going to roam around some of the mining camps and pick up more of the practical side of mining.

With the talks over, it was time to meet Locke in San Francisco. After that I would head for Butte and the doube-jack contests with Jack. Secretly I hoped that John T would be there and that he and I could take a few strokes together, after Jack and I had gotten through. Possibly we might have a turn with three of us, two on the hammers and the third turning.

THE ORPHAN FRACTION

Art and Charlie went down to the station to see me off on the Overland for San Francisco before they returned to Goldfield. Pullman reservations on the train were not available; there were coaches, but I wanted to ride the Pullman. My final decision, when all possibility of securing a Pullman reservation was gone, was to ride the blind. Art and Charlie thought that I was crazy. They were probably right.

The blind on the Overland was not bad. The telescope was wide enough to put my bindle in, in such a position that I could sit on it. The telescope was also wide enough to get into comfortably, more or less, and it was difficult for anyone on the track level to look in to see if some stiff were riding there. Had I not had my bindle along the idea of riding the blind would probably not have occurred to me, and I would have traveled by coach. Art and Charlie stood by to see that no one saw me, and that I was safely aboard.

There was a lot of dust from the roadbed and smoke from the engine, especially when the train started up the grade out of Truckee. By putting a handkerchief over my face I managed to get along all right although it didn't keep cinders out of my eyes. It was hot and stuffy passing through the snowsheds, and I almost choked to death. I had forgotten all about them when I went aboard at Reno. The smoke and steam from the coal-burning locomotive was hot and sulphurous, and the sheds kept it from getting away quickly. There was momentary relief when an opening between sheds was reached, and I could breathe enough fresh air to keep alive. It was not so bad after the summit was reached, and the train started rolling down grade where the speed helped to carry the steam and gas away more quickly.

The train was rolling down grade out of Emigrant Gap where I could see the American River a thousand feet below, and almost a sheer drop to it from the rails. Suddenly the baggage car gave a shudder, and wheels under me left the track and ground along on the road-bed. The coupling of the diner broke free, and cars began to jacknife. The baggage vestibule was hanging over the American River, and I was looking down at it with tops of trees seemingly near enough for me to reach if I put my arm out.

Then the brakes set, the car shuddered again, and I had to use every ounce of strength to keep from tumbling into the canyon below. My bindle had no such luck. It left me, and I watched it tumbling down, rolling over and over as it went on its way, probably not stopping until it reached the river.

As soon as motion ceased, and dust permitted, I was off the blind, milling around among the passengers and train crew that came running to see what had happened. It was not long before a wrecking crew came up to get the derailed cars back on the track. It was a ticklish job, for the diner and the baggage car I had been on were hanging precariously over Blue Canyon of the American River below.

I went down to Colfax on the first work train I could get on. In Colfax I bought clothes, from top to bottom, inside and out, took a hotel room, bathed, and got into my recent purchases. It was not until almost dark that the Overland pulled into Colfax, and when it pulled out some hours later I was aboard as a passenger. A sympathetic porter and Pullman conductor made my passage on the train possible when I added a little something to the regular fare. I spent the night, however, not in a berth, but in the smoking compartment. Anything was all right with me except the blind. I had had enough riding on the blind to satisfy me for life.

After my arrival in San Francisco I went to Locke's hotel, and we were on a train for Bradley that evening. He wanted me to pick up the corners of some of the quicksilver claims that I had located and that he couldn't find. He had a surveyor on the job, and we had the claims surveyed and tied-in in three weeks. I returned to San Francisco on my way to Butte and the double-jack contests. Locke and the stiffs at the mine wished me well and hoped that

Jack and I would win. I told them we hoped to do a good job and let it go at that.

I landed in Butte in time to get in a few days' practice with Jack and to learn the strict rules for the contests that had been established by the miners' union to see that every team had the same chance to win. Jack thought our chances to do well were slim, for there were two classes in the double-jack contests—heavyweight and lightweight. Jack was a heavyweight, and I was a lightweight, therefore we must drill in the heavyweight division. Under such a handicap we would do well to make any showing at all.

I had put on almost ten pounds in the days we were practicing before the contest, and in it, our striking reached 63. Both helped, and we did all right. We drilled two inches less than the champion team from the Calumet-Hecla, and within that two inches were the teams from the Copper Queen and Anaconda, just a little better than Jack and I.

After it was over, we pulled out for Bingham Canyon, where Jack took a job on a Keystone shovel at the Utah Copper. My luck wasn't as good. There was nothing open that I wanted to take, or that would be given to me. All of the assaying, surveying, and draughting jobs had been filled by graduates in mining engineering from the university. There seemed to be no chance for me, but they took my name and promised to let me know when something showed up.

The place was full of stiffs rustling jobs. I looked like a soft kid, and the bosses wanted hard-boiled stiffs who looked capable of putting in a shift. After a week of hustling I got a job that no one else would take—opening hung-up ore chutes. It was not only hard, but dangerous work. It was something I had not done; in fact, I had never heard of that kind of work before, so I took it. At least I would learn something that might come in handy some day. The pay was good—almost double as much as a hard-rock stiff—which was something of an inducement.

A shift boss took me around, showed me the chutes, and pointed out the ones that gave the most trouble and others that gave less. The chutes were in the main drift and two drifts parallel to it. The raises that fed them were large in diameter and made huge rock-walled ore bins. Ore shot into them from the stopes above was held until the chutes were drawn to let it into the ore cars.

The shift boss explained that often huge boulders, or slabs of the blasted-down ore, were broken and fell into the raises. When they did, the raise blocked, and only the ore below them could be drawn until the block was broken. He told me how the block was broken, with sixty powder, but he did not show me. I thought that was strange, but said nothing.

Had I known more about what the work entailed, and the danger, I might have dodged the job, as others who needed the pay more than I had done. Jack called me a fool for not calling it deep enough before I went on shift, but my propensity for learning something new, and getting in trouble learning it, had the upper hand. I reported for work the next morning.

Everything went fine the first two shifts. There were several hung raises, but they were not difficult to break. The third shift was something different. I crawled through a chute gate, above which there was a block, to look things over. The ore was hanging, wedged tight between walls, about fifteen feet above the chute. Two large boulders had tried to get through the opening at the same time and couldn't make it; smaller stuff had wedged them still tighter.

After looking the thing over I crawled out through the chute gate. I selected three of the longest bamboo poles I could find, lashed them together from end to end, strapped twenty sticks of sixty to the small end, put in two number six caps with twenty feet of fuse to each of them. Then I crawled into the raise again. I carefully worked the powder between an opening in the wedged mass, set the lower end of the strapped-together poles, with its long fuse tied on, in a crevice in the wall directly above the chute gate, spit the fuse, and started to crawl out. As the fuse was spit the tommyknockers began to raise hell and instead of crawling out cautiously, I put on a head of steam to get out as fast as I could, caution or no caution.

Just as I made the opening of the chute a dribble of muck started to come down. Then larger pieces, and I was no more than clear of the chute-mouth, and in the drift, when the whole thing let go and came down. I cleared the chute opening, and the chute-set timbers, before the mess hit and took everything out, and before it could bury me. Then all hell broke loose. The powder had come down with the first of the dribbling muck, had lodged on the timbers, been struck by a boulder of ore, and a cap shot the powder prematurely.

The blast tore out whatever of the chute and chute timbers that the falling muck had left, and the drift was loaded with splintered timber and muck. I had managed to sprint a hundred feet from the chute when the powder went, and was lucky to get off with a few cuts and bruises from flying rock. I was sure that I owed my life to the tommyknockers, those unseen, wee, small folk, who came over with the Irish and Cousin Jacks, to tap on the rocks and warn mining stiffs when there is some serious underground danger, as they had warned me when I had spit the fuse.

It took a few shifts to clean out the drift and a few more to retimber and put in a new chute. I wasn't fired, for the same thing had happened to the stiff I had replaced. There was a difference though: he had not got away in time.

Life was too short to be fooling around with that sort of work. I had learned all that I ever cared to know about breaking hung-up chute-raises. It was deep enough, and I went to the office to draw my time. While I waited for the timekeeper to get my check ready, one of the boys at the hiring window told me that a transit man had been hurt the day before. There was an opening for a replacement, in a hurry. I took the job.

My new work advanced me to the engineering staff of the Boston Con. As befitted my new station in life, I moved from the bunk house to a private home on Carr Fork, nearer town and not as close to my work. The job paid one hundred twenty-five dollars a month. The room I had set me back all of $17.50 a month, with everything thrown in. Food was home cooked and wonderful, the landlady did my laundry, darned socks, mended clothes, and sewed on buttons.

Surroundings were congenial, too—the landlady had a pretty daughter. Instead of going down town and spending my evenings at the Old Crow, or one of the other places, I sat in the parlor, talking to the landlady and her husband. Sometimes I drummed on a mandolin that the old man had once played in some glee club. The daughter was an expert on it and taught me how to play. Often after the old folks had gone to bed, we played hearts, putting the mandolin aside. I never had it so good.

Surveying was not new to me, and it didn't take many days before I knew every station in and out of the mine. We were plotting by coordinates and did our own mapping and kept the maps up with the work we did. When the boys found that I could swing a single jack, they let me drill the holes in the mine roof for spuds in which to put the flattened, and drilled, horse-shoe nails that we used as stations. I liked it because it kept me hard and in good shape. No one offered to do the drilling after they found I liked it, so I had enough exercise.

The surveying was by no means routine. Our work took in everything; there was as much on the surface as there was underground. The office miners in Salt Lake did not permit the ordinary tolerances, and the results of our work were to an accuracy that I have not seen since. When we had a particularly difficult piece of work, we could call for a check, with no questions asked, but mostly we rechecked our own work whenever there was any question in our minds as to whether it was within the tolerance allowed.

Larry, another transit man, and I were assigned the chore of surveying for the connection of two tunnels that were being driven from opposite sides of a hill. Work had been started on the results of a rough preliminary survey, and there was a question of its accuracy. The length of the tunnel would be about a mile. There was nothing difficult about it. Larry and I checked each other's surveys. When the ends of the tunnels met, just before Christmas, they were off three inches in grade and four from line.

When Cy Knowles, the super, looked at the connection, he raised our pay five dollars a month, and when Lafe Hanchett, the manager in Salt Lake, learned about it, another five was added. We were not that good, but we didn't object, let the legend ride, and took the raises gratefully, with appropriate modesty.

Winter had set in. It was the coldest in years, and much snow and wind came with it. There was some sort of a deal in the offing. It was rumored that Utah Copper was going to buy the Boston Con property, or consolidate it in that company. Whatever prompted it, the Boston Con management, sitting in nice warm offices in Salt Lake, thought up a good one.

A permanent station and bench-mark was to be established on the top of a hill, from which sights could be taken from every surface station on the Boston Con property. The hill that answered all specifications was five miles distant, and it was a rugged climb to reach the top. Larry and I with our chainmen were the goats selected to do the chore.

After negotiating miles of mountain side with four feet of snow on the level and Heaven knows how deep in drifts, and sometimes breaking through the thin, hard surface crust, we succeeded in bringing a granite block, six inches square and three feet long, to the top. After getting the station in the frozen ground that had to be drilled and shot, we tied in to it from a dozen stations, plotted what we had done, and sent the completed work to the Salt Lake office.

Within two days, back came new instructions from the Salt Lake office. The office miners down there could think up some good ones, sitting in nice warm offices. We were to sight on Polaris from the newly established station. When the time came, it was the coldest, most windy night of the year—twenty below zero and the wind at about forty miles. How we got the job done with hands and fingers so numb we couldn't feel the leveling screws, I never will know, but we did.

When we had worked and plotted our notes, we could not see how we could have avoided an error, for the conditions under which we had worked could easily result in one. There was nothing to indicate that any error had been made, but if there had been we wanted to be the ones to find it, not one of the bunch in the Salt Lake offices. So we set-up and shot old Polaris again. Our work was right both times. If we hadn't earned those raises for making the tunnel connection, we certainly did on those two nights. No one but Larry and I gave it a thought, much less those desk miners in their warm Salt Lake offices.

The pay-off came a few weeks later. The deal between the Utah Copper and Boston Con was still presumed to be in the offing, and the Boston Con claim lines must be run, and tied-in to our station on the mountain top. The survey was a humdinger. The snow was deep, and we had to run lines first to locate each corner, or station, then dig to find them. Sometimes what we had tried to locate wasn't there, so we had to try again. On open ground it would have been so simple it would have been completed in a few days; it took hours a day over a three-month period.

We completed the traverse of the Boston Con, and then we ran out the adjoining lines of other companies, the Old Jordan, Utah Copper, Bingham New Haven, Highland Boy, and some others. As the survey progressed we were aware that something was wrong, but we waited to see what it was until the transit work was completed and reduced to paper.

Near the west branch of the canyon that made Carr Fork was the Utah Copper pit, on the hillside above the pit was the Boston Con with its steam shovel benches and Ben Hur adits One and Two. When our notes were plotted, we discovered that there was a fraction of ten acres here, which neither company owned, nor was there any other owner. The fraction lay about equally on the Utah Copper and the Boston Con workings, and included the Ben Hur adits and a part of the Boston Con shovel benches and a large area of the Utah Copper pit.

We couldn't believe what we had discovered. The two adits were on unowned land and thousands of tons of ore were coming from the unowned part of the Boston Con shovel bench. Utah Copper was mining from a part of a pit that no one owned. If an independent owner had the fraction, he could ask whatever he wanted from either company, or both, for mining ore on land that each of them thought they owned. Or, if the owner wished, he could virtually stop the Boston Con from mining until other exits from its workings were driven, and stop mining from the bench entirely. The Utah Copper pit could be virtually controlled and right of passage could be denied, on the basis of interference with mining operations of the fraction owner.

We possessed knowledge that would be, or could be, blackmail of the highest order, and legal. The same knowledge in the possession of either company, should it come to own the fraction, could result in that company's driving any bargain it cared to propose. It took us ten days to run out the adjoining lines of the Utah Copper and Boston Con, and the lines around the fraction a second time to check our work. The result was the same. The ten acres were still there, homeless, unowned, and unaccounted for, waiting to be located by anyone who wanted to make a fortune.

Larry and I pondered the problem for days. Should we turn in our map showing the claim lines with the orphan ten acres lying between the properties? Should we resign and sometime in the months ahead come back and locate that fraction for ourselves? To make the surveys that resulted in our locating the fraction had taken all of the time we could spare for that work during a three-month period. No one had any reason to suspect that an orphan fraction existed, and if we went down the hill and didn't turn in our notes or the map, the snow would probably keep the Boston Con from running another survey until it melted. Highgrading in

Goldfield was white chips compared to this one! So we decided to turn our notes in to the company.

After the decision was made, I recalled that Lafe Hanchett, who was manager of Boston Con, was an old friend of the family and that I had met him several times since my arrival in Bingham Canyon. He had visited us in New York and probably did not associate the name Crampton with my mother, who had married Sam Rockwell; moreover I was so young when he visited us in New York and in such different surroundings that he never gave the matter any thought. I did not disclose my identity to him when I met him in the canyon.

I also recalled that one summer after she married Rockwell, they had made a trip to Idaho Springs with Fred Himrod and his wife, a former classmate of hers at college, to visit the Hanchetts. Himrod owned two mines near Idaho Springs that Hanchett managed for him before the latter went to Utah for the Boston Con. After reviewing this family history in my own mind, I suggested that we send our notes and maps directly to him in Salt Lake. (I did not, however, tell Larry that my family knew Hanchett.) Larry agreed, and we sent by registered mail our map, notes, a report on the orphan fraction, and a location notice prepared in the name of the Boston Con for Hanchett to sign and record. We told no one in the engineering office of our discovery; all information went directly to the Salt Lake office.

We waited a couple of weeks for some sort of a communication, or something, anything would do, from the Salt Lake office. There was nothing. The void was bigger than that ten-acre fraction. We concluded that the Salt Lake office thought that honesty and virtue was its own reward. But we didn't.

We had been highly commended for a strictly routine piece of work in making the two ends of a short tunnel come together at the right place, and had received raises for it. The job we had done on the claims, and especially the finding of the ten-acre fraction, added up to a great deal more than routine. In addition, we had all but frozen to death establishing the mountain top station. It seemed to us that we had saved the company a fortune, and no comment.

Such ingratitude deserved a just reward. It was deep enough; Larry and I resigned. We hoped that the next surveyors would find another and larger fraction, where it would do the Boston Con a great deal of harm, and locate it for themselves.

It was many years later that I learned from Lafe Hanchett that the fraction was so hot that to disclose its existence was out of the quesion. Somehow the lines were "adjusted," and the fraction remained undisclosed.

Lamertine Mine, Idaho Springs, Colorado, September, 1902. From left: Mrs. S. D. Rockwell (the author's mother), Sam Rockwell, Lafayette (Lafe) Hanchett, Mrs. Fred Himrod. Photo by Fred Himrod. Crampton Collection.

Lafe Hanchett, Bingham Canyon, Utah, winter, 1907-1908. Photo by author. Crampton Collection.

Bingham Canyon, Utah, spring, 1908. Debris from powder-house explosion. Crampto
Collection.

MEN ARE CHEAPER THAN TIMBER

As befitted the occasion, Larry and I went to Salt Lake and celebrated the Boston Con's loss of its two best men. Larry had nothing to interfere with the exuberance of his celebration in leaving an ungrateful and unresponsive company. My celebration, however, was somewhat restrained. The company deserved to lose our valuable services, and I had no regrets for quitting it cold; but the boarding house in Bingham Canyon, and the landlady's daughter, were something else.

When the effects of our celebration wore off, it was time to look for another job. Larry went to Park City while I remained in Salt Lake to see what I could dig up. After a few days I landed a job as mine surveyor, with other duties—keeping progress records on drift and raise contracts, and preparing assay maps.

The mine to which I was to go had been in operation many years and had once been worked with square sets. The company had changed over to cave and fill stopes, and the square-set workings were used only by leasers, who were taking out small lenses of high grade that had been left behind because they were too small to take out on a company operation. I had not worked where mining was done with square sets, and I decided this would be a good chance to learn something new.

When I told the stiffs sitting around the hotel lobby where I was going, they warned me not to take the job. There was nothing indecisive in their warnings. The mine was dangerous. Excepting for the leasers, contractors, and bosses, no white man would work there. The mining stiffs were "bohunks."

I didn't care about who worked in the mine, or what they were, but when I was told that the office miners in Boston didn't give a damn about working stiffs, "men were cheaper than timber," I almost decided not to go. However I had made up my mind to take Sully's advice and learn something new whenever I got the chance. This was going to be my chance to learn square-set mining, and I was going to be a three-month stiff while doing it. I reported for work. I did not remain on the job Sully's three months, however; the mine took care of that.

The management and everything about the mine were all that the stiffs in Salt Lake had said it was, and more. There were more than two hundred men on the payroll, in addition to leasers and contractors, bosses, and the bunch that made up the office force. Not one man on company time seemed to give a damn whether he put in a shift or not, or whether things were right underground or on the outside, except the bohunks who wanted to work and do a job right. No union men were permitted on the job; I got away with it by not letting it be known that I belonged to the IWW and WFM, but I let the unions know where I was, and what went on. The mine was operated in the old tradition that nothing but dollars was of importance.

There was no place in any of the underground workings that didn't look bad. Disorder and confusion prevailed and made a mess of everything in and around the shops and everywhere on the surface, with timber and trash spewed all over the place like jack straws. Even the track was battered up, and wound around, and up and down, on the dump. It was not unusual for an ore car to go over the dump, and sometimes a mucker with it.

Surveys had not been carefully or accurately made, and bad plotting was added to the surveying inaccuracies to make a still greater mess. The results of rotten surveys, bad plotting, and careless mapping would have been amusing had they not been so serious, especially when headings, or raises, failed to connect with workings that they had been intended to meet. As a result, there was poor circulation, and the air was heavy with gas and fumes of long dead powder smoke. The workings, right, left, up, or down, terminated in dead ends, and it was difficult to get around without going miles, it seemed, to reach a face only a couple of hundred feet away. There was only one way out from each of the levels. The Boston office miners were sure doing an expensive mining job in trying to pick up a few extra dollars.

The square set timbers were so rotten that they couldn't take more weight and were ready to go any time. Lagging along the square set corridors was bowed and broken, and gob trickled through the openings and fell to the floor or tracks, which resembled a roller coaster. It took a full-time mucker to keep the tracks cleared so that ore cars could be run over them. A few leasers were taking out high grade, using square sets to hold the ground, and tramming through old corridors. The leasers were the only ones on the job who blasted light. They had one man in addition to the mucker cleaning tracks and constantly watching for signs of the old square sets and timbers coming in on them. They were making good money, or they would not have been there. Those square set stopes were one hell of a place to work, even for the high grade the leasers were getting.

The adit tunnel to the new workings had been badly driven. It looked as if the shift bosses had encouraged machine men to drill their holes deep and load and tamp them with powder. With the exception of some sound ground in the drift, a few hundred feet in from the main adit mouth, everything had been badly shaken by overloaded holes. There was very little timber in the adit tunnel and less in the mine workings. The roof and walls were continually sluffing, sometimes catching a man and hurting him badly. Frequently men were killed in cave-ins or sluffs. The tommyknockers were overworked and had to put on double shifts.

After I had looked the place over I knew what the stiffs in Salt Lake meant when they told me, "down there men are cheaper than timber." Sully was right when he said that I would learn something new at every mine. On that job I learned the most important lessons of my life: how not to operate a mine, and how not to treat or handle men.

There was no organization in the office, shops, or any of the underground workings. The miners were bohunks, fresh from the old country, and couldn't speak much English. They didn't know much about mining, and no one took the trouble to teach them, but they did the best they knew how, learned fast, and were all good hard-rock stiffs. The contractors and leasers were old and experienced hard-rock stiffs who had grown careless in the confusion around which they worked. The shift bosses let everybody get away with anything. The stiffs were even permitted to shoot their rounds whenever they had finished drilling and loading.

One afternoon, not long after lunch, I was measuring the advances made by each contractor and had gotten the faces of all but one. This contractor had three faces—a drift with two cross-cuts. The drift was about one hundred feet ahead of the cross-cuts, each of which had been driven about fifty feet. I had passed the cross-cuts on my way to the drift face when I smelled smoke from lighted fuse. I should have known something was wrong when I did not see lights or hear men working. Like everyone else in the mine, I had grown careless. I didn't pay attention to the busy tommy-knockers either. I had gotten so used to their warnings that to heed them would have meant doing no work at all.

After passing the cross cuts on my way to the face of the drift, I crawled between the wall and an empty ore car, and stopped to call to anyone who might be nearby. There was no answer. At that moment a hole in one of the cross-cuts blasted and my candle went out. In a split second I had turned the ore car on its side and thrown myself into it. Then all hell broke loose. All three faces were blasting at the same time, the round in the drift and the rounds in the two cross-cuts. As usual the holes had been overloaded and rock from the drift face was catapulted against the bottom of the overturned car. Rock was thrown across the drift in both directions from the cross-cuts.

When I was sure that the holes had blasted, I got out of the car and, without light, crawled on my hands and knees to the main drift five hundred feet away, through powder smoke that was strangling and that almost made me tear my lungs out from coughing.

I laid in the drift a long time, breathing a little cleaner although smoke-filled air, while I waited for strength to come back so that my legs would carry me to daylight. The blasts had so deafened me that I did not hear an ore car coming down the drift, nor did the mucker know I was laying across the track until the wheels of the car struck me. By some miracle performed by a guardian angel, my legs were not broken, but they were badly cut and torn. When the car was on the track again, the trammer threw me on it with the other muck and trammed me out of the mine and into daylight and fresh air.

It took days to regain my strength and for my legs to heal enough to get around on them again. I was to get over the leg-healing hurdle, but the concussions from so many close-by blasts had damaged an ear drum beyond repair. I should have called it deep enough, but I was as stubborn as a balky mule. I was not going to quit until I had learned everything that I could about square-set mining, in three months.

While I was laid up with my injured legs, the super hired another surveyor, thinking I would draw my time when I got around again. I fooled him and the new man, Steve Kelly, a swell stiff, was assigned to me as a surveying assistant. Had I known I was to be docked for the days I was laid up, I would have quit and told the super that it was deep enough, then drawn my time; but pay day did not come until ten days after I went on the job again. By the time I found out I was to be docked, it was too late for me to do anything about it. Nature later made the decision for me.

After my having nearly been blown to pieces, the office gave orders to blast only at lunch and tally, and a man was to stand-by to give warning that a round was being fired. My close call had accomplished something, but there were a lot of other things that needed attention, particularly the storage and handling of powder.

The powder magazine was a half-mile from the tunnel entrance. In it were stored over a hundred tons of powder, thousands of blasting caps, and miles of fuse. Powder and capped fuse for blasting were issued from a powder house fifty feet from the mouth of the adit. Behind the powder house was a twenty-foot tunnel in which two hundred cases of powder were stored regularly, and with it caps and fuse. The powder house and storage tunnel were too close to the adit, and with all of the carelessness prevailing, anything could happen.

Three days after I was back on the job, with my legs still unhealed and in bandages, the powder house and tunnel storage behind it blew up and took the tool house and blacksmith shop with them. The powder monkey, nippers, and hard-rock stiffs who were drawing powder for the shift-end rounds were killed. How it happened there was no way to learn, and the number of men blown up was learned only by checking the payroll. There was not enough left of the blasted men from which a count could be made.

While the mess was being cleaned up, work was being rushed to save the adit tunnel, which had been badly damaged by the blast and was caving. In addition to the hell caused by the blast, the snow had been melting slowly, and the water from it seeped into the ground instead of running off. There was not only damage to the tunnel mouth, but water was pouring through the cracks and faults opened by the blast, and the ground above it was working. There were a lot of bad sluffs and caves, and the stiffs were having a tough time timbering to catch up and keep the tunnel from caving. Had it been timbered when it was driven, there would have been less trouble, but not having much timber made work doubly difficult and dangerous. The tommyknockers were not having a holiday.

After four days of hard and dangerous work, it appeared that the worst was over and that the new timbers would hold. When Steve and I came on for the graveyard shift, the worst of the caving ground had been caught and there was not much sluffing, even where the roof was still open and new timbers were to go in. Bob, the Cousin Jack shift boss, thought the tunnel safe enough to do timbering from both ends, and when Steve and I agreed with him, we got together a crew of seventeen bohunks and started into the mine to timber from the four hundred foot station toward the tunnel mouth. We should have known better than to take a chance, although there was little or no sluffing, for those tommyknockers were putting on the biggest show of their lives.

At the six hundred station were two wide cross-cuts where tools, lagging, and timber were stored to be used when heavy snow made it difficult to get at the timber piles and to the shops outside. The place was in disorder, as was everything else, but the idea was sound and practical. In one of the cross-cuts were open lockers for stiffs to hang clothes, and shelves on which to leave lunch buckets. After shedding coats and putting lunch buckets away, we started the bohunks to work on framing timber for tunnel sets. This done, Steve, Bob, and I went down the drift to look things over before commencing the job of timbering.

As we were going down the drift, we heard yells coming from the tunnel and saw men running toward the mouth. In a few seconds there came the sounds of breaking timbers, then a grinding rumble, followed by a depressing silence and a violent rush of air as if a big blast had gone, or an over-pressured air line had broken. Our carbide lamps went out but we did not need light to know that the tunnel had caved.

Our carbide lamps lighted again, we looked at each other and, without a word, headed back to where we had left the bohunks. By the time we got to the cross-cuts the bohunks had their candles going again and I could size them up. Fear was written on every face, as I know it must have been on mine, but there was no sign of panic. All of us knew we were in for trouble, and there was not one who wasn't scared stiff. We were caught behind a tunnel full of rock that would take a lot of days to muck out and timber. A caved-in adit, with no other exit from the workings, was bad—rotten surveying and management had taken care of that.

There was almost nothing we could do to help. The little that we might do would waste our strength, and we were going to need all we had. Relays of hard-rock stiffs outside, working in twenty-minute shifts, would break through, sooner or later, and it was up to us to last it out if we could; our chances were much better if we did nothing.

There was no immediate danger where we were. We could hear the ground working and still coming down into the tunnel. Caving had stopped at the drift end where we were; a few feet more might sluff, but even if that happened, only a few tons would come, and it would not block the tunnel any more than it already was. The workings had a lot of bad ground but none between the four hundred and the fifteen hundred foot stations; we would keep away from the bad ground deeper into the mine unless we were driven back by water.

The workings were heavy with dead powder smoke and the smell of water-soaked timber. With thousands of feet of drifts and cross-cuts in the workings, it might be some time before the danger from mine gasses would become serious. Water dripping from the roof and coming out of the walls would furnish us more than we needed to drink. Although the water was a blessing in one way, its constant dripping from above would keep our clothing soaked. It would give us a lot of trouble, moreover, if it did not drain through the caved ground and backed up into the workings. This danger was somewhat offset by the grade of the drift, which was steep (about four inches each hundred feet toward the mouth of the tunnel).

Food was our greatest problem. When we took inventory of the graveyard lunches, which had not been eaten, we found that the lunch buckets contained less than we had expected. But we knew that with water men could last much longer than with food and no water.

Candles and snuffs were collected and put in an empty powder case. Knives, candle holders, and miscellaneous junk were deposited in the owners' lunch buckets, to be claimed later. We all kept our watches, money, plug tobacco, snoose, and matches. When the inventory and collection were over, the lunch buckets and candles were cached down the drift where they would not be a temptation and could be guarded against raids.

We were going to be in darkness long before the stiffs working in the tunnel could reach us. By burning one candle at a time there would be enough to give light less than four days. The carbide lamps that Steve, Bob, and I carried could be kept lighted a few hours with the unburned carbide in them and with the reserve we carried with us in tins. We hoped to get everything prepared for the wait ahead before resorting to candles.

Our first consideration was to make our waiting place as comfortable as we could while we were fresh. The intersection of cross-cuts and the drift at the twelve-hundred station was selected as the safest place. The ground was good and there was no danger from sluffs or caves. Moreover, the level of the drift floor here was over three feet higher than at the closed end of the caved-in tunnel. There were more than three miles of workings below the twelve-hundred station, and they would hold a hell of a lot of water. It would take a long time before it reached us or became deep enough to drive us back to the far end of the workings where the ground was loose and dangerous.

The bohunks were divided into three groups, one of five and two of six. Lots were drawn to determine who would occupy each of the three locations. The five group drew the drift, and the six groups drew the cross-cuts. Steve, Bob, and I had already selected the drift below the intersection.

When we realized what we were up against, we had wasted no time worrying and had gotten down to cases in a hurry. Everything was under control, the food and light situations were taken care of, the location of our new homes decided, and there was nothing left but to start building. While Steve and Bob were getting the home work under way, I went down the drift to size things up and to take a good look at the cave-in.

When I reached the five-hundred station, I paced the distance to the caved ground. It was eighty feet. Four hundred twenty feet of a caved-in tunnel lay between us and daylight. If all of the tunnel had caved, it would take no less than three weeks before the stiffs on the relays of rescue workers could break through. It would be longer if mucking was difficult. It would also depend on whether a lot of spiling had to be done, and how much work it would take to get sets in to hold the ground. The stiffs would not do a fancy job, a hole to get through would be enough, but they would make sure that the hole would not close behind them. It could be more than three weeks—a lot longer than I cared to think about. It might be less if some of the timber put in during the past few days had held. The thought was encouraging, but I didn't dare let that hope take over.

Faint sounds came through the caved ground, or the rock walls. I didn't know which, and it didn't matter. The sounds told me that the stiffs were on their way. I would have known it anyway, even had I heard no sounds. They were giving it everything they had. The sounds were a reminder that we had sent no signals, and it suddenly dawned on me that a message must be gotten to the stiffs outside to let them know that we were alive.

As I knelt to send a message on the compressed-air line, I realized that there was no pond of water under me; water coming down the drift was flowing through openings in the caved ground. It was a welcome discovery. Then I signalled—tap-tap, a pause, tap-tap-tap—and hoped the compressed-air line was not so badly broken that a message would not go through. Then I waited. There was an answer—tap-tap, a pause, tap-tap-tap—faintly. From the sound I knew the line had been badly bashed and damaged but messages would go through. Then, to let the stiffs know we were all alive I signalled "twenty," and when the reply came I got up and went back to where the bohunks were working.

Work on our new homes went fast. Lagging and timber from the cross-cuts was used without cutting, and a couple of ore cars made it easy to get it to the job without manhandling. Our homes were to be platforms, one in each cross-cut and two in the drift, one above and one below the intersection. Barriers erected at the end of each platform made a hollow square of the intersection and we hoped it would discourage the bohunks from wandering around and visiting. The location of our platform on the lower side of the hollow square at the intersection, with the barriers between us and the bohunks, gave control of almost any situation—panic, demoralization, or a rush for food. We did not misjudge the bohunks, we did not expect trouble, and there was never any sign of it, but anything could have happened under the increasing tensions of passing hours and days.

Toilets were built in a corner at the end of each platform and against its barrier, with empty powder cases used as thunder-mugs. Burned carbide from a pile in the cross-cut near the lockers was used as a deodorant. The bohunks got fresh drinking water by dipping it from the trench along the tracks, before the water went under the platforms. Our water was taken from the eleven-hundred station cross-cut, much below our platform and away from anything that might flow down the drift. From the survey station in the roof of the intersection we fastened wire and hung an empty cap-box to the lower end. A lighted candle in the cap-box threw light into the drift and the cross-cuts.

It was six hours from the time of the cave-in before everything was completed, and then there was nothing more to do but wait. Before sending the bohunks to their platform homes, we all had a piece of sandwich meat and some bread. Somehow, while the work of home-building had been going on, three of the bohunks had fashioned crude musical instruments. Two were made from powder cases to which necks had been attached, and each had four strings made from tightly drawn rawhide shoe laces, taken from the bohunk's high work boots. The third was a flute, fashioned from a piece of pipe with a four inch long crack. One end was plugged and the other had a very creditable mouthpiece.

The bohunks began making a holiday of the situation. The groups would take turns chanting to the deep bass notes of the powder-case instruments, or to the high-pitched flute. When a chant ended, peals of laughter would follow, punctuated by comments passed between one group and another. We couldn't understand much, but what we did was enough to leave no question that it was profane and ribald.

Steve, Bob, and I joined the festivities as best we could. Sometimes the chants went on for a long time. Then there would be quiet until some of the bohunks became restless or could not sleep, and started it all over again It was good that the bohunks had found something to do, and their antics broke the monotony, for them as well as for us. It was not easy to wait and do nothing.

Dragging hours were not made shorter by the routine of giving signals on the compressed-air line. Each succeeding hour interval passed slower than the preceding one. Watching the candle to see that it burned to its bottom, and no more, then waiting for the split-second to put in a new one, lighted quickly as the burning candle flickered out, was a strained diversion. It made us more aware of time than did the signalling. When half of the candles had been burned, we knew that darkness would come sooner than we had expected. The candles were cheap and of an inferior grade and burned rapidly. Oncoming darkness was nothing to look forward to.

At first we doled out food in small portions, but the allowances were increased when we found that the meat and bread were beginning to go bad. The muggy-warm dampness and the gasses from the dead powder smoke were no doubt responsible; both had already begun to affect us also. On the second day, what was left of the food was divided and eaten. The food did more harm than good: it was in worse condition than we had suspected, although there was no odor of spoilage. After it was down a short time, all of us began to retch, and we lost it all. The retching continued for hours. It would have been better had we thrown it away; strength was lost to each of us for having eaten it.

Every hour Steve, Bob, and I took alternate turns going down the drift to signal and to listen for the sounds of work that seemed to be getting closer. At first there had been no blasting, and the sounds seemed to be coming rapidly closer. On the second day there were sounds of machine drilling, then, after a short silence, a few shots. As time passed, the machines drilled longer as the sounds came closer, and the shots were heavier. The longer drilling and heavier shooting were discouraging. It was not difficult to guess what was happening. The stiffs working to get to us, had passed the loose broken ground near the tunnel mouth, and were hitting large blocks of caved rock as they came closer. The blocks must be huge, or the blasting would not have been so heavy; unless there was a damned good reason, the blasting would be held to light-loaded holes. Our guess of three weeks to break through was raised. Only by a miracle, we thought, could the stiffs get to us in time.

The third day was less than half over when an extra heavy shot sent a concussion through the workings that put out our lighted candle. Before that no one had given a thought to matches, no one smoked in the mine at any time; we chewed plug or used snoose. The only time matches were used was to light lamps or candles and that had not been necessary except immediately after the cave-in. Every stiff carried a few in a pocket or in some battered-up metal case. All were useless. The water in our water-soaked clothing had ruined the loose ones, and dampness had gotten into the metal containers where others were kept. There was no way to get a candle going again, or a snuff, and two shifts of light remained in them if they could be burned.

Waiting with light was bad enough. Without it, everything was changed. There was less conversation, and the music and chants from the bohunk's quarters were heard less frequently, then not at all. The darkness made us realize our trouble acutely. When we had light, what we were up against seemed as if it were an unreal dream; without light, it was a nightmare and far from unreal.

During the time we had had light our ears had become accustomed to sounds of the stiffs working to get to us through the caved tunnel, of men turning and groaning, and of heavy breathing and snoring, all expected and therefore unnoticed. In the darkness a new sound was heard. Whether any of us had noticed the ticking of watches, all twenty of them, I don't know, probably as unnoticed as other sounds we had become accustomed to. But with darkness, sounds from our watches, irregular and at random, were as loud as a boiler factory in which a drum corps was practicing.

Not many hours of darkness passed before I had stood the new noise as long as I could. I crawled over the barricade and told the bohunks to hand over their watches. Not one of them hesitated. No doubt the noise had been as unbearable to them as it had been to me.

When I crawled back to our platform, with the boiler factory going full blast in my pockets, Steve and Bob handed me their timepieces without a word. Then I crawled down the drift to the eleven-hundred cross-cuts, felt the floor for a hole deep enough with water to cover the watches and drown the noise, found one, and dumped the mess into it. I made my way back to our platform and dozed off, in the silence of usual and noticed sounds. The boiler-factory drum-corps situation was solved.

Our clothing, clear to the skin, was waterlogged. The continual dripping of water from the roof could not be stopped, and there was nothing in the workings with which to make a shelter to turn it away from us. When we lay in some position for a while, the part of our bodies not facing the mine roof would warm up, and the water-soaked clothing also. When we turned again, which was often, the dripping water from the roof would chill us again in a matter of moments. It was as if we were being boiled in cold water.

Water-soaked clothing was softening otherwise hard flesh all over my body, and the flesh became so loosened that it would slip and crawl over the meat underneath whenever I moved. In addition, my injured legs were swelling and becoming increasingly painful. It was not my legs, though, that gave the most trouble: it was the flesh loosened from the meat and muscle underneath. Changing from one position to another became impossible, as impossible as it was to remain in one position longer than a few minutes at a time.

My teeth were chattering almost constantly, as if I had malaria, a result of the wet, cold, water-logged clothing. My jaws were so tired and sore that I tied a bandanna around my head and chin to keep them from moving, and to stop the agony. It didn't stop the jaws in their effort to chatter; the bandanna loosened and had to be retied time after time, until finally, from sheer and hopeless weariness, I gave up trying and let them chatter as they would.

Occasionally the movement of my jaws would stop. Whenever it did, I could hear the chattering of the teeth of the stiffs around me. They were having the same trouble as I was, both from their cold-boiled flesh and chattering teeth. I was sorry for them, but I was glad to know that I was not the only one who was having the same kind of troubles.

Weakness and exhaustion were overtaking us long before it should. Water-logged clothing and bad air that was getting worse were bad enough, much too bad, but the spoiled food was getting in its best work. We had plenty of water, but only a few had held it down since the food was eaten. It took several days before all of us were over that jolt. By the time we were all again able to hold the water we drank, our weakness was so great that nerves, already strung to high pitch, had all but reached the breaking point.

We had to crawl down the drift to the caved-in ground to give our signals to the stiffs outside. At first we had walked, feeling our way along the drift walls, but that was impossible now. Even crawling was becoming more difficult. The flesh over my knees was so loose that I had to take most of my body weight on my hands, since they had not softened up as much as the flesh that was covered by clothing. We could not signal as we had before, at regular intervals. There was no way to tell time except by the few moments of almost silence from the direction of the tunnel when a fresh relay took over at its twenty-minute relief interval of rescue work. We tried to keep track of the changes, but concentrating and keeping count were impossible. Our minds were getting a little fogged, or we would doze and miss the count. We signalled whenever the next whose turn it was felt strong enough.

The second time it came my turn to signal, after the light was blown out, I signalled and, as usual, waited for the answer. When it came, I could not believe what I had heard. I signalled again, giving my code, four quick and closely spaced taps at the end. Then I waited, my heart pounding my ribs out and my lungs refusing to breathe. It was a long, long wait. Then it came: tap-tap, a pause, tap-tap-tap, followed by two quick taps. It was John T! And before I could answer there came another, this time with three quick taps at the end. Sully was there too!

My heart pounded so hard, and I was so exhausted by the waiting and listening, that I couldn't return the signal right away. What I had heard was our code signature signal to each other, a hold-over from the days of Cripple Creek. Finally, when my strength returned, I signalled again, ending with four quick taps. Then I leaned against the drift wall and cried like a baby. It was a long time before I had control of myself again.

On my way back to the platform, Steve and Bob met me about half way. They had become frightened and thought something had happened to me. When I told them about John T and Sully, their hopes rose as had mine. We knew there were two of the best hard-rock stiffs in the country outside, and that if they had not already taken over to boss the job they would do it now. Jack, whom I knew was already there, would help see to it that the three of them did.

In the hope that we expressed to each other, we secretly knew that not even John T, Sully, and Jack, could get one more ounce of effort from the hard-rock stiffs than they were giving already. Nor could they give more themselves. Whether for me or anyone else, it would be the same, every ounce. Who was inside didn't matter, we were hard-rock stiffs like themselves, and they knew what it must be like to be inside waiting.

It must have been on the seventh or eighth day that the sounds of work and drilling sounded closer, and the shots were lighter. It was a strange and sudden change. Had our minds not been fogged, we would have known what had happened, and a lot of strain would have been lessened. All of us were well on the downhill side, and there was not one of us who did not know it, but we kept our thoughts to ourselves. The bad food, reeking wet, gas-filled air, and cold-boiled flesh were getting in a lot of fast work.

I have no way of knowing when it happened, but it must have been well along on the ninth day. There was a piercing shriek. It came from the bohunk's home in the drift. Then unintelligible words. After that a few moments of silence and a dozen voices rose to question what was going on. Then another shriek, and sounds of a man running, the sound of a body falling, or hitting something, and another shriek followed by deep moans. More sounds of running, more shrieks, more sounds as of a body falling or hitting something. All the time the sounds grew dimmer and dimmer; finally, a scream that pierced the workings as would the whistle of a freight engine. Then silence.

I waited for more sounds to follow, hoping they might, and yet hoping that none would come. Nothing came. The silence was deadly. No one spoke, there was nothing that one could say, and nothing anyone could do; but for a long time breathing sounds drowned out whatever other sounds we had become accustomed to hear. There was not one of us who did not know what had happened. One of the stiffs had come to the limit of strain, and his nerves and mind had broken. He had gotten up, started to run, and beaten himself when falling to the floor, or against drift walls, time after time. There was little chance that he had come through alive.

We gave up going down to signal on the compressed-air line after that; fear pinned all of us to the platforms. I could not have gone even had I wanted to. There was no strength left to go, and my legs were so swollen that the pain was unbearable. I put the bandanna over my mouth and tied it behind my head to help keep from making audible sounds. I wanted to moan, or to cry; it didn't matter which, I wanted no one to hear whatever it might be. Neither Steve nor Bob moved either. The only sounds I heard were those of heavy breathing, and moans as men turned from one position to another. All other sounds seemed to have ceased.

I was in a dazed and nightmarish doze when the break-through was made. A sudden rush of ice-cold air brought me out of it, momentarily, and I could hear voices down the drift, but they did not seem to get nearer, and I went off into unconsciousness again. The next I knew was when I heard the voice of John T, shouting over and over, "Cover up your eyes, here we come with lights."

There was no chance of my putting anything over my eyes. I was so weak I couldn't move. I do not think any of the other stiffs could either. Then, as the stiffs who had worked to get to us came closer, we knew it was not a dream. And then I heard sobs, deep-breathing sobs, trying to be held back, but breaking out nonetheless, for the strength to hold them back was gone. Then I was sobbing, too, and tears running down and smarting my tender-skinned face.

The next thing I realized was that John T, Sully, and Jack were bending over and putting me on a stretcher, or something. They wouldn't let anyone else touch me, excepting a doctor who felt my pulse and shot something into my arm. After that I vaguely remember being carried in darkness through muggy air into the fresh outdoors. I tried to wipe my eyes to see, but Sully clamped his hand over the bandage that he or John T, or maybe Jack, had put there, and I gave up. As I pulled my arm down, John T gave me the information that it was night and that I wouldn't be able to see anyway.

After I had been wrapped in blankets that seemed to weigh a ton, I was carried down the hill and put into what I thought must have been a buckboard, the way it bounced around. A little later I heard the chugging of a railroad engine and knew that there was a train waiting to take us to Salt Lake, or somewhere, to a hospital.

I wondered how many of us there were left to come out, and asked the boys. "All but one," John T told me. Then he added "Shut up" and I knew that the "one" was the poor stiff who had cracked up not many hours before the break-through came. There were nurses and a doctor who took over when the boys got me onto the train, and I got a little more attention than any of the rest. I was the first to be brought down and reached the train minutes before any of the others arrived.

It took weeks to recover, not so much from the nerve strain and semi-starvation, as from my legs, which had become badly infected. And also the cold-boiled flesh took a long time to get settled again and tie itself to the meat that would hold it in place. I heard a doctor tell a nurse that he did not understand how I had come through with my legs in the shape they were.

John T and Sully came to the hospital three times a day and never left until they were put out. Jack stuck around a week before he returned to Bingham Canyon. He had gotten to the mine a day after the cave-in. John T had been in Jerome when word of the cave-in reached him. Sully was in Telluride. Both drew their time and were on their way. Each had brought a couple of hard-rock stiffs with him. They were "champs," they said. But there were other champs, too, every hard-rock mining stiff of them from all over the mining country, who came in, just because there were hard-rock stiffs in trouble and who needed help. The boys told me that we were lucky, as if I didn't know it, but when they let me know that more than two hundred feet of tunnel had been held by the new timber, I knew more about how lucky we were.

But I didn't know it all, until later. The powder blast had loosened the ground not only over the new workings where I had been, but also over the older part of the mine where the square-set stopes were. The old and rotten square-set timbers let go, and that part of the mine had caved, too. Many hard-rock stiffs had been caught in the collapsing drifts and corridors. A few had been brought out alive, but others not so lucky had been dug from under the mass of broken rock, gob, and crushed timbers. There were a lot more still buried, and it would take days of dangerous work before all of the buried and mangled bodies were recovered.

John T and Sully didn't know where the other was, or that the other was coming, until they met on the same train out of Salt Lake, on their last lap to the mine. They had gotten there on the fourth day and had spent six days helping to dig us out. When they reached the mine, no one knew in what part of it I was working, although some thought it was in the new workings. They decided to start helping there and to give our Cripple Creek code sign-off signals whenever signals came from inside the caved workings. They were about to give up to join the square-set relief crews when my signal came through. That was their story, that is all I ever got out of them, there was nothing else to be said.

THE BATTLE OF THE CENTURY

The doctor did not dismiss me from the hospital for almost a month, and then ordered me to remain in Salt Lake where he could watch my legs and take care of them until they were healed. The idea of sitting around doing nothing but look after me did not make a hit with John T or Sully. Nor did it appeal to me, either, but I was weak and did not care to use my legs more than was necessary. It was as hard on me as it was on them. However, they were not going to pull out until they were certain I would not get impatient and go off to do some fool thing again before I had recovered.

Neither John T nor Sully had any confidence that I would not get into trouble, wherever I went. Getting lost in the mountains, any fool would have known enough to hobble the animals. Riding the blind through the snow sheds and nearly being thrown into Blue Canyon when the train derailed: no bindle stiff would ride the blind as I had, since it was double luck that I lived through the smoke and gas, let alone not having been thrown into Blue Canyon when the train derailed. And almost caught in a caving raise: only damn fools and morons took jobs like breaking chutes.

There was a lot of sputtering and fuming as John T and Sully went over the list of my crimes, but nothing compared to the steam they let off when they got around to the six weeks I had put in on the last job. They were irrational. Every stiff in the country knew that mine. The miners' unions had been trying to do something about it for years. I should have taken the advice of the stiffs in Salt Lake and not gone there! Every excuse I could offer was beaten down, I even tried to lay the blame on Sully and his advice to get all of the experience I could, and not remain at a mine more than three or four months. Sully drilled and blasted a couple of rounds in that one and left me with nothing but waste to work on.

Running into a lot of blasts was something that the boys could not understand. Why hadn't I paid more attention to what they had taught me in Cripple Creek? I couldn't get past that one. Laying across the tracks, I could have laid in the drainage ditch along the track. The cold water would have washed some of the dumbness out of me. If I had been in a decent mine, the mucker would have been pushing the car so fast my legs would have been cut off; lucky I was in a loafer's mine. But to get caught in the cave-in! Didn't I know that before a big cave the ground stopped moving for a while, getting ready to go all at once? And what about the tommyknockers, were they not on the job, or didn't I hear them? There was no answer that I could give that would satisfy the boys—so I let them sputter it out. They were right.

But it wasn't the trouble that I had gotten into that almost broke my pick with John T. It was that I had drilled in the double-jack contests in Butte without him. Why hadn't I let him know? As if I could. Jack Commerford was a good hard-rock stiff, he had known him for years, but he had some English blood in him somewhere, and that spoiled any Irishman. John T was all Irish. That English blood in Jack was what caused us to lose. John T knew that he and I could have beaten those Michigan stiffs. But what John T forgot was that the blood in me, and in Jack, combined, would make one complete Irishman and one Englishman. I guess that John T, knowing I had no love for the British, thought I was Irish too.

Another thing that troubled John T was that it was getting along toward the Fourth of July contests to be held in Butte, and here I was, fresh out of a hospital, with legs that wouldn't hold me up, and I couldn't get in shape in time. The only consolation was, in the mind of John T, that next year we would go up and show those four-flushers how to really put a steel through granite.

In the meantime, while the boys were waiting, the company had generously sent my time check, delivered by the timekeeper, whom I didn't like. When he told me that he had talked the manager into breaking a rule in sending me the money, instead of my going to the mine to get it, I thought better of him. One look at the check changed my mind. I didn't know which I disliked the most, the company or the timekeeper. There was no point of compromise, neither was any good.

The check did not include time for the days I had been laid up with injured legs, after my bout with the blasts. The timekeeper explained that it was only carelessness that had gotten me into trouble. The company never paid anyone for being hurt because of carelessness. He had a point there. Someone had been careless all right—the company, the contractors, the trammer, and myself. The company held the winning hand, there was nothing I could draw to win. It was an unbeatable four aces.

Included in the check was time for the week when I returned to work after the blasts affair. I thought that one over, carefully, and was at the point of sending that time back to the company. I had earned nothing that week. My legs were so bad that I could not get around much and, whenever no one was looking, I took five. There was no one to see what I did, most of the time, and I got a lot of rest. Sully jumped me on that one, and it didn't take much talking for me to change my mind.

There was no pay for the ten days we were caught underground by the cave-in. The office miners no doubt figured that we were taking time out. No one was paid for not putting in a shift. We had put in enough shifts in those ten days to last a lifetime. We had not worked, I couldn't argue that, but the shift we did put in was a hell of a one, ten days wrapped into one long package.

Nothing was provided for payment of the hospital and the doctors. The timekeeper told me it was I who was sick, not the company, and let it go at that. I started to spit a short fuse and blast him all over the place, but John T and Sully told me to keep my shirt on and to forget it. Forget it? Never! I hoped those office miners in Boston would have nightmares the rest of their lives, and hear that poor stiff running down the drift, screaming, and beating himself to death as he ran.

When he left, the timekeeper told me there was to be no charge for the train trip from the mine to the hospital. John T, Sully, and Jack had taken care of the first part, the trip from the mine to the buckboard; the buckboard owner had charged nothing, for me or for the others carried in it later that night. I should have known that the company wouldn't have had anything to do with anything that cost money.

John T and Sully took me from the hospital to a hotel when I was released. After they were certain I would be all right, John T went back to Jerome, and took Sully along, to put him on the shift he was bossing in the fire zone. John T said that he was going to steam off some of the weight Sully had put on while he was waiting for me to get well.

It was another three weeks, after John T and Sully had gone, before the doctor dismissed me and told me to get going. I wrote the boys that as soon as I landed somewhere, I would let them know where it was. With my writing shift in, and purchases of a new work outfit and some store clothes, I went to Ogden and caught the Overland for San Francisco. I had vague ideas of going to the quicksilver and remaining there, with Locke, for a few months, while I got my strength back.

As the Overland was approaching the causeway over Great Salt Lake, I drifted to the platform of the observation car and found C. W. sitting placidly among the dust and cinders that swirled around him. He was glad to see me and immediately wanted to hear all about the cave-in. He had heard about it, but wanted the story from me. When I had finished telling what had happened during my six weeks at the mine, C. W. was furious. He knew the company and some of its officers; they would hear from him, personally.

C. W. was making a trip to see what he could dig up for New York clients who thought mines grew on trees, and that ore could be had for the asking. C. W. hoped that he could find something for them. There was no limit to time or money, or where he went to look for the mines. Find one was all that his backers wanted. I was glad when he asked me to go along with him.

We covered a lot of the mining camps of northern Nevada and looked at a few in the Mother Lode country and then drifted over to see Locke. The quicksilver mine was doing well and a battery of retorts was turning out streams of quicksilver. Locke knew quicksilver mines and advised his backers to sell, which he did for them. In a few weeks Locke was going to leave for Mexico where there was a silver mine he thought good enough for his backers to take over and operate.

After a month drifting around California mining camps, we headed for Kingman. C. W. was going to see if he could find something in Arizona. We pulled into Kingman in time for the

Fourth of July celebration. After it was over, we rented a buckboard and went out to look over mines in the nearby camps.

C. W. found what he wanted near Chloride. The mine had been butchered by inexperienced operators looking for high-grade, but there was a lot of good milling ore showing, and it looked as if it had the makings of a mine. C. W. asked me to remain to get the camp set up for a small operation and the mine cleaned up while he went back to New York and Boston to get his backers lined up on the financing.

The mine was not a long walk from the Chloride post office, and I spent evenings in the saloons playing twenty-one or swapping lies with the Arizona hard-rock stiffs. There was more social life in Chloride than any place I had been, and it was not long before some of the married stiffs I met in saloons were asking me to their homes for supper or Sunday dinners. Life really picked up, though, when I met the daughter of a mine foreman, a cute little blond, who made me forget the landlady's daughter in Bingham Canyon.

Things around the mine picked up, too, and there was a great amount of unexpected work to be done that kept me out of the office. I wrote C. W. that I needed help. If he knew of a man in the East who could fill the job, please send him out, otherwise I would put on one of the local talent. C. W. replied that he was sending a man from New York who wanted to learn the mining "game." C. W. parenthesized the word "game," and it worried me. The man would arrive a few days after I received the letter, and it would not be necessary for me to meet him. When he arrived in Kingman, he would come out on the stage to Chloride on his own.

Late one afternoon, a few days later, the Kingman stage pulled up in front of my office and disgorged something. It was followed by a glittering array of walrus hide suitcases, hand bags, and a trunk large enough to hold a wardrobe for every, and any, occasion. The trunk was so heavy that it made the stage driver grunt when he threw it off, and he made no attempt to lift or touch it after the chore of throwing was completed.

The disgorged something was wearing a straw hat with a red and yellow band. From the hat hung a slender black silk cord, with gold trimmings, that dropped to his shoulder and was tied through the lapel of his coat. The suit he wore was a beautifully fitting masterpiece, a tailored affair, with pants knife-edge creased. His linen shirt had a detachable collar, so high his neck stretched. The red tie with black stripes was tied for an evening stroll on Picadilly or Fifth Avenue. It seemed to be one of those British school ties. Beau Brummel had nothing on this apparition.

I was glad my office was at the mine and not in town where all of the stiffs could see the thing and give me the razzing of my life. It was something for the birds, and looked as if it had stepped from the pages of a fashion magazine, or of something pictured in one of Siegel Cooper's advertisements—of what would be around when you "meet me at the fountain."

The apparition advanced to the office door, introduced itself, handed me a letter from C. W., and waited. It was ill at ease, for the moment, shifting weight from one foot to another as if preparing to take off on some intricate dance steps. I had been prepared to meet a tenderfoot, but not this. After I had read the letter of introduction from C. W., which was noncommittal and vague, I told him what his duties would be. He was not interested and changed the subject to things more dear to his heart—his clothing and travels.

The bright red tie with black stripes had been bought in London, as had been his suit, made by some lordly tailor "by appointment to His Majesty." Black patent-leather shoes were from Berlin, socks from Vienna, underwear and other personal things, from Paris. His hat, made in some Latin American country, was from Madrid. His shirt, bought in New York, the only American-made thing he had on, was of linen made in Ireland.

Then came a dissertation on travels. The apparition had been everywhere in Europe and recited his wanderings as if memorized from Cook's travelogues. He was a cosmopolitan, all right, his clothes and everything about him. No one at the time, in Chloride, or Kingman, to which town he went on Sundays to attend church services, were given a chance to forget it. Within a matter of twenty-four hours someone had nicknamed him Sweet Caporal. It fitted, and it stuck.

It took a great deal of effort to get Sweet Caporal into clothing suited to his coming status at the mine. He insisted on wearing only what he had brought with him, most of which was on a par with what he wore on his arrival. To get him into digging clothes,

I assigned him only work that would soil or damage what he wore, beyond repair or redemption. That did it.

When Sweet Caporal decided to don digging clothes, I went with him on the stage to Kingman to be outfitted. A broad-brimmed Stetson was his first purchase. After that came high laced-boots, corduroy pants, a heavy blue wool shirt, and a red bandanna to use as a tie. He insisted that he would stick to Parisian underwear. I had no objection. No one could see it. His final purchase was several pair of soft leather gloves, to keep his hands clean.

It wasn't long before Sweet Caporal was going into town to stand up to the bars and drink raw whiskey, with whiskey chasers, as if they were water. It did not phase him much more than water did; he was a good two-fisted drinker and could hold it. It was not long before Sweet Caporal began thinking of himself as "one of the boys." The stiffs were getting tired of listening to Sweet Caporal talk about himself and his travels. No one was interested in hearing the same things retold, night after night. There was complete agreement that he should be put in his place. First it was a mission to get a sky-hook. After that he was sent to get a nipper, a left-handed monkey wrench, and other errands as obviously impossible. He never tumbled to the fact that the stiffs were pulling tricks. No one was discouraged, for all of Chloride had agreed it was necessary to make a man of Sweet Caporal.

It was not long before there was the suggestion of a badger fight. Several nights of elaboration over the drinks Sweet Caporal was buying made him eager to see one. He was told that it was hard to find a badger, and dangerous to make a capture when one was found. No one offered to find one until Sweet Caporal became so insistent that there was nothing to do but to get one, someway. He was going to see a badger fight, even if he had to go alone to make the capture. No one let him get away with that idea. If he went alone, he would probably be killed; the least he could look forward to was to be terribly maimed and crippled. There was no more comment on his making the capture single-handed.

There was nothing left to do but go into the hills, risk the danger of serious injury, possibly death, from the claws and teeth of the badger, and capture one. There were long discussions of who would try to capture one of the beasts. It took several days before a couple of stiffs agreed to go. They insisted it was Sweet Caporal's idea in the first place, and he must go with them. A couple of men with rifles were to go along. If things became too hot, the badger would be killed by them.

There were four in the party who went out with Sweet Caporal one night when the moon was full. With them went a like number of "frontier men," with rifles. When the party reached a place where it was known that badgers had been found on previous and similar occasions, the party split, half going up one canyon and the rest another. Both parties were to return to the ore wagon that had taken them out, in not more than an hour. If one of the parties did not show up on the dot, the other would go to the rescue.

The teamster remained with the wagon and horses, to quiet them if a badger came along and frightened the animals. He too had a rifle. On the wagon was a cage, with bars of quarter-inch steel. The cage door was small and opened and closed on a vertical slide. The cage was in a specially prepared wooden box, reinforced with steel bars on the outside. The illusion of the badger, if captured, as being dangerous was greatly increased by this masterpiece.

By the time Sweet Caporal and his party returned to the wagon without a badger, the other party was there, and in the cage was a badger. It had been trapped the night before and staked out near the spot where the wagon stopped, and was in the cage within minutes after Sweet Caporal and his party went out to capture one. Tall tales were told of how close the stiffs had been to being clawed and bitten in making the capture. Sweet Caporal left no room to doubt that he was glad it was not his party that had caught the brute.

Back in Chloride there was a great commotion when the badger was brought in. Drinks were freely bought for the party that had made the capture, Sweet Caporal paying for most of them. Sweet Caporal went to the cage, often, to look at the badger, but every time he did he was brushed aside by someone else who wanted to take a look, usually a long one. Every expert on badgers in the house said it was the most vicious they had ever seen. There was no one who disagreed. Sweet Caporal was so excited, and had gotten such fleeting glimpses of the animal, that he would not have known if it were a lion or a kitten.

The stage was set for the final act. The fight was scheduled to take place a few days after the badger was captured, and on a flat a quarter-mile from Chloride. There was a lot of room for the battle ground and spectators.

During the days before the fight there were great excitement and many discussions whenever Sweet Caporal was within hearing distance. Odds on the fight shifted frequently, and these were posted on a large bulletin board outside one of the saloons. There was always a crowd around when Sweet Caporal was nearby, and thousands of dollars in "windy" bets were made, in voices so loud he could not fail to hear.

Try as hard as he might, Sweet Caporal couldn't get a dollar of his money covered. He wanted to bet on the badger. No one could accuse him of not trying. Everyone he asked had already bet all of his loose money. It was tough luck. Why not go into Kingman to see if he could find anyone there who would take a bet? He did. Again more tough luck. No bets. The build-up was going fine.

Sunday had been selected as the day for the fight. Almost everyone worked the week around, but would take a day off if there was something special, or important, going on. The badger fight was one of the important things; no one was going to miss it. The crowd assembled at the battle ground, and, by three o'clock, the time the fight was scheduled, there was not a place left from which a good view could be had. There were a score of deputy sheriffs, with six-guns loosened in their holsters and ready for any emergency. They had been carefully selected and deputized for the occasion.

Everything was ready. The cage, with the badger in it, was at one end of the battle arena, a space one-hundred feet wide and twice that long. Men stood on either side with dogs held by chains; there were two on each side, the dogs straining at their leashes and barking at they knew not what. There were scores of dogs in reserve, behind the roped-off arena. All were held by chains in the hands of some stiff who would let them loose when the order to do so was given.

The sheriff had been chosen master of ceremonies. He was strictly "neutral," although, he too, had made bets, whether on the dogs or badger he would not tell. When the sheriff went to the center of the arena to put the affair in motion, he looked around and called for time out. He ordered that anything over a foot high be brought out for the women and children, among the spectators, to stand on.

Tables and chairs from the eating places, card tables, boxes, and whatever else would answer the purpose, were brought out and arranged around the arena. Women and children were mounted on these improvised stands and the sheriff went to the center of the battle ground again to look things over. After carefully rearranging some of the elevated onlookers, he declared that everything was satisfactory. He was taking no chances that any of the weaker sex or children would be attacked and possibly injured by the badger. He was very particular to make the point clear to the onlookers.

In a booming voice the sheriff announced the rules of the contest. The rope that led to the badger's cage was to be pulled, whereupon the stiff pulling it would run to the sidelines and join the onlookers. As soon as he was clear, two dogs were to be unleashed, one from each side.

As the badger disposed of each pair of dogs, another pair were to be released, and so on until the sheriff declared a winner. Under no circumstances was the fight to be permitted to continue for more than thirty minutes.

Everything was ready. The sheriff looked everything over carefully, on all four sides, and called for the stiff who had been selected to pull the rope to come forward to release the badger. No one came. There were catcalls and threats against the person who had been selected. Someone surely had been. Was he too frightened to come forward and perform his duty? The coward! Tar and feather him! The situation was acute. Would the fight come off? Uneasiness pervaded the crowd, and the sheriff was having a hell of a time keeping it under control. The culprits in the crowd who had instigated the thing were having a gay time keeping the near riot going.

The sheriff let the thing go on as long as he dared. Then he called for a volunteer. No one came forward, and again the crowd was calling "coward, lynch him," alternated by "tar and feather him." The sheriff raised his arms and asked for quiet. He had a solution. Was there anyone who had not made a bet? There was —Sweet Caporal, although to give the devil his due, he was still trying. The day was saved. Sweet Caporal was told to step into the arena. He didn't want to, but he did, on legs and knees that were obviously shaking.

Sweet Caporal didn't have much choice. He could refuse, be hooted down as a coward, stripped, tarred and feathered, and run out of town on a rail. He might even be lynched. Or he could pull the rope, take his chances with the vicious badger, hope the dogs would stop it while he was running to the sidelines, or pray the badger would be shot before it reached him and tore him to ribbons.

The choice between the devil and the deep-blue-sea had nothing on the decision Sweet Caporal had to make. He did the only thing that he could—chose between the lesser of two evils. The crowd cheered and went wild.

The sheriff stationed Sweet Caporal within a few feet of the cage. He handed Sweet Caporal the rope leading to the door through which the badger would come, and he grasped it firmly in his trembling hand. He looked at the crowd around him, as a man about to be lynched might do, and at their faces, friendly, but serious. No one seemed pleased that he was to be the subject of a Roman holiday and be fed to the lions.

The sheriff's voice boomed last minute instructions, and it echoed back from the nearby hills. If any trouble developed, the spectators were to remain where they were, they were not to move, he and his deputies would see that no one was injured. Nothing was mentioned about Sweet Caporal standing in the hot sun, shaking with face as pale as the linen handkerchief he was using to mop his sweat-bubbled face.

I walked to the side of the blond and waited. She took a dim view of the whole thing. It was a dirty trick to play on an innocent, unsuspecting boy! I laughed. More than that there wasn't anything funny about it. I should be ashamed of myself! I wasn't. I shouldn't let this thing go on! How was I to stop it? Sweet Caporal had brought it on himself. Sweet Caporal was an employee of mine, it was disgraceful, what would C. W. think? I knew what C. W. would think: give him the works. As to being funny, little did she know, nor I, even: it wasn't funny, it was hilarious. When it was over, she thought so, too.

The sheriff asked if everyone was ready. No answer. He asked for quiet and fired his six-gun. The show was on, the crowd let out a roar. The dogs on the sidelines and others behind the lines, in reserve, barked harder than before. I took a glance at Sweet Caporal, and he was trembling so hard I feared he would faint, but he controlled himself quickly.

Sweet Caporal pulled the rope, and ran with it tightly clutched in his hand. He was too frightened to let go. Then he looked back. There was the badger, six feet behind, following as he ran and making the strange, hollow-like, metallic sounds of a badger in hot pursuit. His speed increased, if that were possible; he was already making a dash that would beat the hundred-yard record.

Something had gone wrong with the role the dogs were supposed to play. None had been released. Sweet Caporal was alone with a vicious badger, in the middle of that wide open space. It was not gaining on him, but he didn't gain on it. His speed couldn't be increased, but the record pace continued.

Sweet Caporal had not bargained on fighting a vicious badger single-handed. The dogs were supposed to take care of the fighting part. In terrorized haste he stumbled and fell, at the foot of one of the few trees in Chloride. Sweet Caporal took another look at the badger, still six feet behind, and climbed the tree.

He still had the rope in his hand; even while he climbed he had been unable to loosen his hold on it, his fright had been so great. As he climbed, the badger climbed too. When a branch some eight feet above the ground was reached, Sweet Caporal straddled it, and looked down. There was the badger, two feet from the ground, the gunny-sack that had been wrapped around it had become loose and was falling off. As it fell, there was exposed, on the end of the rope, dangling as a pendulum, a large white-enamel thunder-mug, the late afternoon sun shining on it, making it plain for all to see!

Before Sweet Caporal reached the limb of the tree, howls of laughter, shouts, and cat-calls filled the air. But they were nothing to the sounds that echoed from the hills when the dogs were released, as he threw his leg astraddle of the limb. Every last dog, barking like mad, made a line for the tree. Then, beneath the tree, looking up at Sweet Caporal, they let out barks and howls, or fought among themselves. The savagery of the din sent shivers along my spine. What it was doing to Sweet Caporal no one knew. He had let the rope, that had held the badger, drop from his hands, and he hung to the tree trunk and limb for dear life.

Instead of laughing it off, or at least trying to, and waving clasped hands at the crowd, he let out a stream of profanity and condemned us to hell, and added to the insult by voicing disparaging remarks about family ancestry. That did it.

No one attempted to call off the dogs. They barked and fought among themselves until an hour later when some stiffs, who had gotten enough drinks to make them feel sorry for him, went out to lend whatever assistance they could. They drove the dogs off and helped Sweet Caporal down. He should have gone into town and bought drinks, and had he done so all would have been forgiven, in spite of his vehemence when he discovered he had been tricked. He had been the object of a wonderful show, everybody was in the mood to forgive anything, or anybody, even Sweet Caporal. He did not appear.

When the stage pulled out for Kingman the next morning, Sweet Caporal was on it. He was dressed as he had been the day he arrived. It was his parting gesture of contempt. He was the same conceited, arrogant cosmopolite that had drifted into camp weeks before. He left his digging clothes on a fire in back of his quarters.

A few days after Sweet Caporal reached New York, an express delivery was made to his door. In a box, with holes to admit fresh air, was the badger "thunder-mug," and the badger, alive. C. W. wrote that Sweet Caporal was not appreciative.

CHICKEN-FEED

After the badger fight things around Chloride returned to normal. Sweet Caporal had created unexpected diversion and a great deal of entertainment. It had cost a lot of time from work, but it had been worth it. Everybody was sorry to see him go, not that they wanted him around, but because a snipe hunt with raiding Indians on the war path had been planned that would have been a dilly. As if the badger fight had not been enough! Anything that might follow would have been anti-climax.

November rolled around, but C. W. had not returned as he had planned. The holidays were coming up, and I knew he would not be out again until they were over. It would be well into the New Year before he got back. Since September, when C. W. left, the hole had passed from prospect to a mine.

Art Clark had written that the Goldfield business had about folded up. I wrote him to sell the office building, pack the assay and survey equipment, and bring it with him. He pulled in a few days after Jack Commerford, to whom I had written at the same time. John T was my first choice, but he was pulling down more at Jerome than he would in Chloride, so I dismissed the idea. He would have come, and Sully too, had I asked. Jack was as good an all around hard-rock stiff as was John T, so it made no difference; my preference for John T was more a matter of sentiment.

When Jack arrived, I put on three shifts. Art took over the assaying and engineering. Another bunkhouse was built, and an addition to the mess hall and kitchen. The shaft was being sunk another two hundred feet, and more faces were gotten under way developing ore. Three stopes were already putting out shipping ore, and additional raises for stopes, on newly developed high-grade, were increasing the paying production. Before C. W. returned, there would be two maximum-tonnage cars going out every week.

A week before Christmas I went to Kingman for a holiday fling and shacked up at the Hotel Beale. The blond and her mother came in the following day, to visit friends and to remain for Christmas. Jack decided that Chloride would be good enough and that he would not come until Christmas day. Art was going to be married to a girl he had met in Tonopah. She was coming to Kingman, and they were to be married Christmas afternoon, then to go to Los Angeles for their honeymoon.

Kingman was busting at the seams. Every mining stiff who had a stake and could get off was in town, as were ranchers, with their families, from miles around. The few strangers were conspicuous, but were made welcome by the townspeople, mining stiffs, and ranchers, who knew each other so well that they called each other by first names.

Christmas trees were everywhere; the largest, in the Beale lobby, was loaded with candles, decorations, and gifts for the hotel guests and for friends of anyone who wanted to put one on, to be distributed when the time came. The candles were lighted on Christmas eve, after the gifts had been handed around, and several stiffs stood by, with buckets filled with water, to put out a fire should one of the lighted candles on the tree start a blaze.

The Beale barroom was the most decorated in town. Even the gambling tables had sprigs of holly. Mistletoe hung from the ceiling. It was purely decorative. The only women who went into the place were girls from the line, whose restrictions limiting them from leaving their segregated district had been lifted for the holidays. It made no difference to them if there were mistletoe or not.

There was an unexpected limitation to visits with my blond. She was not allowed out of the house after supper, and her mother made me leave as soon as I had helped finish washing the supper dishes. Daytime hours, however, left nothing to be desired. Every day, during the week the blond was in town, I rented a buckboard, the blond put up lunches to take along, and we drove around in the hills looking at the scenery. We stopped whenever the horses seemed tired. Far be it from us to overwork them! It was rough and hilly country, and the horses tired easily and often. I do not remember much about the scenery.

I had the day shift well taken care of, but the night shift was different. There was plenty to do in Kingman, but the variety was limited to hotel lobbies, bars and their gambling tables, and the line. There was no place in Kingman that I missed; if there was it was deep underground, and I couldn't locate the shaft going down to it. I was getting a vacation all right, but not much rest. Bed didn't see me until it was about time to get up and take the blond for another buckboard ride.

Christmas came with its exchange of presents, then Art's wedding, early in the afternoon. The wedding breakfast was prepared by the blond's mother and served at her friend's home. After we had gotten Art and his bride on the train for Los Angeles and their honeymoon, Jack took the mother and my blond back to Chloride. My blond cried a little at the wedding and wished that someone would ask her to marry him. Months afterwards it dawned on me that she was accepting a proposal, if I made one. I do not know what I would have done had I realized, at the time, what her tears and remark meant.

I spent most of Christmas evening in the Beale Bar playing roulette, with less than indifferent success. I had gone through a big grubstake and was down to all but a few hundred dollars and ready to call it deep enough. I changed my mind, though, and decided I might as well be broke as the way I was. Get it over and go back to the mine the next morning, a week before I had planned.

I had watched the board in the hours I had been mucking money over the waste dump. Nineteen was the only number that had not come up. Whether it was a hunch, the law of averages and percentages, or just dumb luck, I do not know. Whatever it was it brought home the bacon. I made nineteen the key to my last bet, and my last dollar. A hundred on the number, and fifty each on the column, dozen and color, and as the ball was about to be spun I got another fifty on the quarter. It was about as stupid a thing as I could think of to do, but the agony would be over suddenly and temptation completely removed. The ivory fell in the nineteen slot.

The crowd around the wheel, as well as the shills and girls from the line, let out a whoop; then there was silence. The dealer counted out my winnings on each bet and shoved them over. My winnings were paid, mostly in five and ten dollar gold pieces, with a sprinkling of twenties; there were stacks of silver dollars. There was chicken-feed, too. The dealer was far from happy. He had let me bet double the house limit on a number. He had to get even with me in some way. The dealer called to the bartenders and told them to send all of the dimes, nickels, and pennies they had. When the chicken-feed delivery was made, the dealer counted it out and shoved it over to me. There were a little over fifty dollars in the pile. It was most impressive in appearance, but it sure spoiled the looks of the stacks of gold and silver.

There were plenty of offers to help me cart the stuff away. It was heavy, and an Irish buggy would have helped. Some thoughtful stiff handed me two canvas ore-sample sacks to put the stuff in, but it was not the dealer behind the wheel! I put the gold pieces and silver dollars in one sack, and the chicken-feed in the other. Then I walked over to the bar and called for everyone to come up and have a drink. The girls from the line followed me over; they stood near the bar and formed a half circle about me as I stood and waited for all to get the drinks they had ordered. Most of them were whiskey, with whiskey chasers, at two-bits a throw for the combination.

I put the two bags containing my winnings on the mahogany, and waited. The drinkers stood six deep in front of the bar, and it was some time before everybody in the crowd had drinks in their hands. None could drink until everyone had been served, and the buyer had given a toast, sometimes appropriate to the occasion, sometimes not. But a toast must be given. It was Beale bar tradition.

While I was waiting, a stranger, whom I had noticed at the roulette wheel during the time I was playing it, edged through the group of girls and came to rest beside me. When I was certain that everybody had their drinks, I raised mine and toasted: "Here's to crime, may prostitution flourish." The girls let out a whoop, then the crowd downed their drinks and drifted back to resume gambling or whatever else they had been doing.

The girls hung around the bar and wanted me to spend some of my winnings on them, any one of them, take my pick. The only thing I could think of that I wanted to do, at the moment, was to give each of them a ten dollar gold piece, as a Christmas present, and to buy them a round of drinks. As I was ordering for the girls, the stranger who had edged in beside me included himself as one of the crowd with the girls.

After the round was downed, I started to settle for what I owed and dove into the sack containing the gold pieces and silver dollars. The stranger put his hand on my arm and suggested, in a loud voice, that I should pay in the chicken-feed that the dealer and bartenders had wished on me when I was paid my roulette wheel winnings. The bartender did not like the idea, but I dumped the stuff on the bar and watched him while he counted out the amount that I owed. There was still some chicken-feed left after the bill was paid, but not much, and I dumped it back into the sack.

As I turned to leave the bar and go to the lobby to put my winnings in the hotel office safe, the stranger asked if I would sell him what was left of the chicken-feed for twenty dollars. I told him there was not that much left, but he said he liked the feel of chicken-feed. He passed over a twenty gold piece, which I accepted, and I turned the sack over to him.

Meanwhile, the bartender, who had been an observer of the transaction and had overheard the conversation about the chicken-feed, put his pick in. Anyone who wanted chicken-feed badly enough to pay twice what it was worth was crazy. He thought he had seen everything, but this was the top limit of stupidity. His comments were uncomplimentary, but the stranger grinned, asked the girls still standing around to have a drink with him, and passed it off.

When I returned from the hotel lobby after depositing my winnings, having first taken enough out to last the rest of the night, I returned to the bar. The wheel that had given me my winnings was not crowded, so I went back and started playing again. My luck continued to be good, but I did not press it and was satisfied with modest winnings.

After a while I tired of the wheel and drifted to the bar to have a few with some stiffs that I knew. The second drink was barely down when there was a whoop from the crowd and cheers from the girls standing around one of the wheels. I looked over and saw the stranger preparing to gather in a killing. He too had bet a number straight-up, with side bets, and won. Christmas did not appear to be the night for roulette wheels to do their best for the house.

As soon as the stranger had collected his winnings, he made for the bar, with all of the girls in the place following. He spotted me, and the bartender who did not like chicken-feed, and headed in our direction We made way for him to reach the bar. He pounded the mahogany, first with one fist, then with the other, alternating with banging the sack of chicken-feed I had sold him earlier. As he pounded he called for everyone to come up and have a drink.

The crowd took little time reaching the bar and stood deeper in front of it than when I had bought for the house earlier. Newcomers who had been attracted by the commotion when I had set them up were still around, hoping, but not expecting, something to happen again. They were rewarded for their infinite patience.

The drinks were handed out, mostly whiskey with whiskey chaser. Then silence while waiting for the toast that must always come. One did:

> Here's to California
> Her rivers are dry
> Her clouds without rain
> Her woman lack virtue
> Her men have no brain.

It was a variation of the hated toast, often given on the other side of the Colorado River, with California substituted for Arizona. No one seemed to have thought of the substitution before, and it brought the house down.

As soon as the stranger was told what he owed, he fingered the sack with the small change I had sold to him, and asked the bartender if he objected to being paid in chicken-feed. The bartender took a look at the familiar sack, waved his arms, and, in a voice loud with disgust, replied that chicken-feed would be all right. There was no question that the bartender didn't like the idea but, inasmuch as there was nothing he could do about it, would surrender and accept it. The conversation between the stranger had been so loud that it sounded as if a fight were coming up. Those who had been leaving the bar stopped moving, and many came back and crowded around to see what it was all about. No one wanted to be too far away to miss a good fight.

As the bartender turned to serve some stiffs a few feet down the bar, the stranger put a hand in a pants pocket and brought it out, filled with cracked corn. This he put on the bar, sorted it, arranged it in rows and fingered it as if he were counting. He then smoothed it out and made one pile of it, reached into his other pants pocket and brought out another handful of cracked corn. The same counting performance was repeated, with the bartender watching as if he could not believe what he saw.

Then, bringing the cracked corn together to make a second pile, the stranger remarked that it was still not enough and again went into a pocket and brought out a few grains, looked it over, and commented that was it. While this new wrinkle in paying a bar bill was taking place, the girls around the stranger and the stiffs who could see what was going on passed the word to those who could not see, and there was a tenseness of expectation.

Not a word had been spoken by anyone other than the stranger, and that only as to the sufficiency of the chicken-feed. The bartender, whose face was getting red with anger, was undecided what to do and too astonished to think of anything to say. It was better for him that way. Suddenly the silence was broken by shouts of laughter, yells, and cat-calls. The house was in an uproar. Everyone was on his feet in a split second, ready to fight, or to run, or do whatever conscience dictated.

The girls were in hysterics, and the stiffs at the bar were pounding their fists and yelling. It wasn't a fight and wasn't going to be one, but something had happened and the crowd lost no time doing all it could to find what it was. Most of them had to be told. Meantime large numbers were coming in from outside to see what was going on, and the place was so crowded that no more could come in. The curious, who had to remain outdoors, had the word relayed to them.

The bartender pounded the bar with his fists, then with a beer-barrel spigot. It made no impression. The crowd became more boisterous. The bartender finally grabbed a bung-starter and pounded on the bar so hard that dents were made in the mahogany. The noise let up. At least the voice of the bartender could be heard when he mounted a beer keg, behind the bar, to let the crowd have it.

He had told this stiff, pointing to the stranger, that he could pay his bar bill with chicken-feed. Never had such a low, scurvy, snake-like trick been played on him before. This stiff, again pointing to the stranger, was a miserable skunk, the lowest and most unspeakable of all characters. No horse thief or cattle rustler had ever reached the depths of dishonor, perfidy, or treachery, as had the stranger. What had the stiff done? Why this low-down miserable, contemptible thing had paid in chicken-feed! That's what he had done! And it was the best cracked-corn that Arizona ever produced!

Anyone who didn't believe it could take a look. The stiff was a stranger. He didn't know right from wrong. He didn't know up from down, so he wouldn't call the law, or ask the crowd to lynch him. Hanging was too good for him! "Drinks for everyone, on the house, come and get it!"

The din that arose when the bartender was through compared to nothing less than the cheers for Admiral Dewey as he marched up Fifth Avenue after his return to New York from his victory at Manila Bay. It took almost as long to serve the gang that had crowded into the Beale Bar and hotel lobby as the Dewey parade had lasted. Every stiff in Kingman who could walk, came in, sooner or later—later if they had to wait outside for someone to leave. There was no way for the bartenders to know who had been inside, or who had come in after the invitation to drink on the house had been given. There was no way to know who should, or shouldn't have a drink, so all got one, and more than a few had their share of seconds and thirds. There were very few bar sales made during the rest of the night.

The house had had a bad night—two big losses on roulette wheels, and drinks for almost every male in Kingman and about all of the girls from the line. But it was Christmas and who cared? The bartender and the house were good sports that night; it was seldom they were not, but this night was the prize of them all.

When the place had quieted down (it didn't get back to normal that night), the cracked corn was put on a small platform, under the large central mirror behind the bar, for everyone to see. Behind it, neatly printed, was a sign: "We do not need chicken-feed. We have more than we can use." The sign and cracked corn lay there for a long time, until rats finally got it.

The chicken-feed idea looked so good to me that the next time I was in San Francisco I tried it at the M. and M. It worked. The bartender I tried it on in Jim Jeffries place, in Los Angeles, called the cops, but they did not get there soon enough. I paid the bill before they arrived and never tried the stunt again. The old west was on its way out.

There were only a few in the Beale bar the evening after Christmas. Things had quieted down and everyone was starting to rest up for the New Year celebrations. There were only a few playing at the gambling tables, and the shills were having a hell of a time trying to coax a few drunks to go over to one of the tables and part with their money.

More than the usual number of girls from the line were in the barroom, not trying to get business, but because it was the holiday season and the restrictions that would send them back to the joints would be in effect again on New Year's day. They were going to make the most of the little freedom that was permitted them while they could. Some were very good dancers, and I took turns around the small floor with quite a few.

The "professor" who had been at the piano during the week was exhausted. He had had to stay on the job, tired or not, and was more than glad when I offered to spell him at the piano. The girls and stiffs seemed to like my playing, and it netted me free drinks whenever I became thirsty. The coins tossed to me for playing some tune, or for no reason at all, were put in a tin can and given to the professor when I quit for the night and the professor took over again. He had done all right, as he did every evening, or night, that he played.

About the time I was about to call it a shift and head for the hay, Estelle, a madam from the line who ran the biggest parlor house, came to the piano and asked me if I would come to her place and play for a few days. She told me that the professor who played regularly had gotten into a brawl the night before and the law had taken him over. I could think of worse ways to kill evenings for the next few days and told her I would be down in the morning to see her, after I had had time to think it over.

Before Estelle left, we sat at the bar. When a few drinks had mellowed the atmosphere, she unburdened her woes. She was a "trouper" and had come west with a stock company that played one-week stands in towns and camps, wherever there was an "opera house." Estelle played the lead in the *Still Alarm, Because She Loved Him So, Old Homestead, White Heather,* and *Uncle Tom's Cabin,* favorites wherever the troupe went. Everything was going fine until the manager skipped out with the troupe's money and the blonde who played juvenile parts. The men of the cast scattered and left the girls to shift for themselves, in the only business in which they could earn enough money to get them back to New York.

Estelle and her girls hated what they were doing and the tunes the regular professors pounded out. They longed for something different, more like the songs and music popular in New York. My turn as professor in the Beale Bar convinced Estelle that I was her man.

On one of my trips to New York, not long afterwards, I drifted into Voll's Music Hall on One Hundred and Twenty-Fifth Street for an evening of relaxation. On the platform that was called a "stage" were Estelle and her girls, going through a well-appreciated routine. As the number ended with the customary flourishes of the girls showing their wares, Estelle spotted me at a table close to the stage. When the applause and encores were over, Estelle and all of the girls rushed to the front of the stage and called for me to come up.

Apparently I was too slow getting under way, for Estelle and two of the girls ran down the steps and dragged me from the table onto the stage. When I reached the center of the platform, Estelle put her arms around me and gave a kiss that brought a roar of appreciation from the house, and one that I will always remember. Then raising her arms for silence, she introduced me as the "best professor of piano music west of the Mississippi." She added that I had "tickled the ivories" in a place she ran in a "rip-snorting, mining camp in Arizona." When I left her place, she and her girls couldn't take it any more, and ran for New York.

Estelle told the crowd that she knew I would follow her wherever she went, and here I was to prove it. "Give them a tune, Frank." Not long before coming to New York I had been in Los Angeles and had drifted down to Venice and the "Ship Cafe." The professor there, and his partner, had just written a new song, "Casey Jones," and introduced it to the crowd in the "Ship Cafe" that night. I played it for Estelle and her audience, probably the first time it had been heard in New York, and the crowd went wild. After I had finished playing, Estelle asked me to write the words, and in a few minutes she and her girls were singing them as I played. I played others, too, and the girls sang, too; but they did not make the hit that "Casey Jones" did. I went to Voll's almost every night after that, until I pulled out for the West again.

The next morning I headed for the line and the house in which Estelle wanted me to be the substitute piano "professor." No one was up yet; it was not noon. So I went over to the piano and started playing. I drummed out some Christmas songs and a few others appropriate to the holidays, and was down to enjoying myself thoroughly in the quiet surroundings when I heard humming and the words of a song coming from somewhere behind me. I turned to see a dozen girls in their robes; it was they who were humming and singing. My thoughts had been on the songs and playing them; the surroundings and the place I was in had completely gone from my mind.

More girls came in, then Estelle, and they formed a half circle and asked me to keep on playing. I asked what they wanted me to play. Anything but the things the professor usually played. They wanted the familiar Christmas and holiday tunes, or anything that was decent.

The girls of the line were not very busy over the Christmas holidays. It was a time when their thoughts ran deep, and they became introspective and sad. I played the things they wanted, and as I played they sang, softly, and forgot, for a little while, where they were. I was glad that I had come.

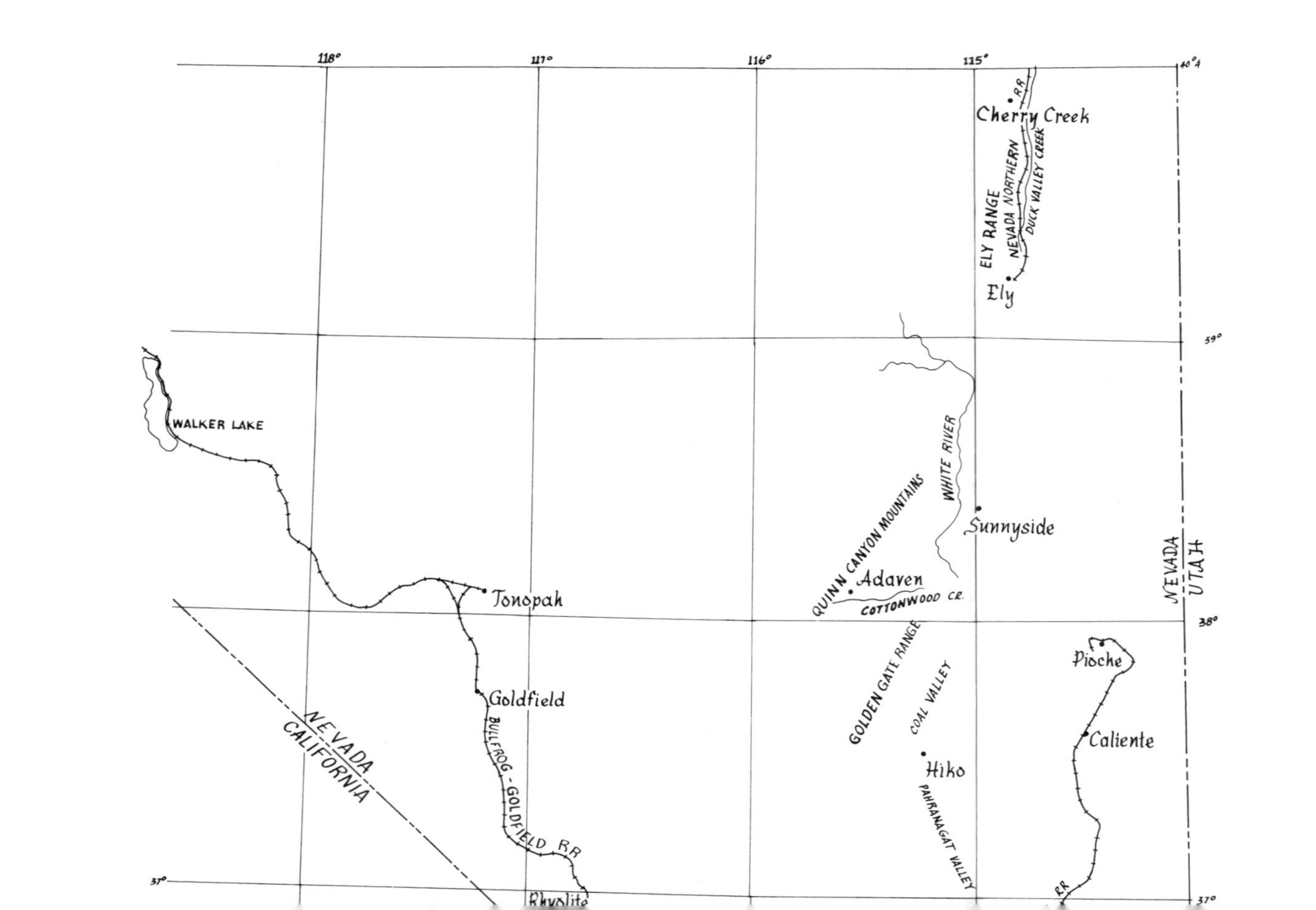

118°
117°
116°
115°
40°
39°
38°
37°
RR
Cherry Creek
ELY RANGE
NEVADA NORTHERN
DUCK VALLEY CREEK
Ely
WALKER LAKE
WHITE RIVER
Sunnyside
QUINN CANYON MOUNTAINS
Adaven
COTTONWOOD CR.
NEVADA
UTAH
Tonopah
Goldfield
BULLFROG - GOLDFIELD RR
NEVADA
CALIFORNIA
GOLDEN GATE RANGE
COAL VALLEY
Hiko
PAHRANAGAT VALLEY
Pioche
Caliente
RR

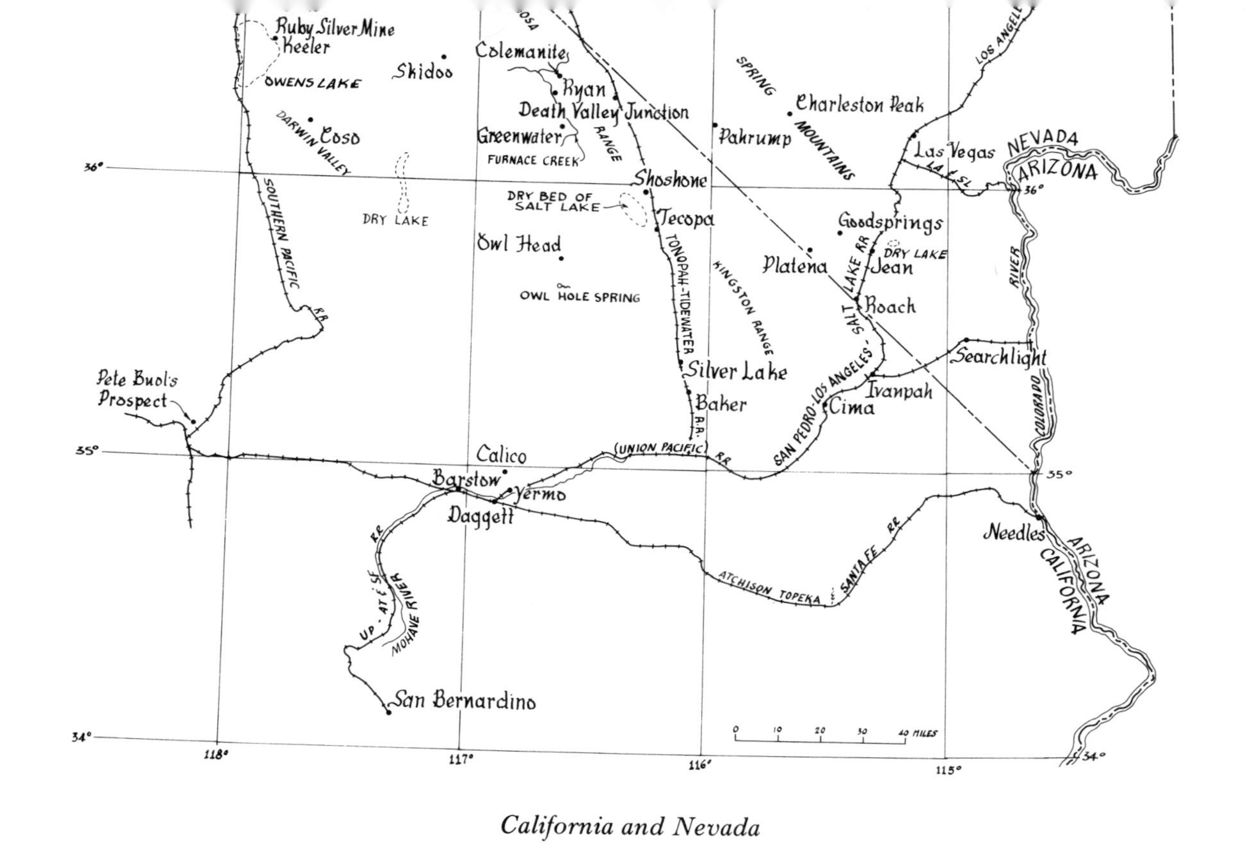

California and Nevada

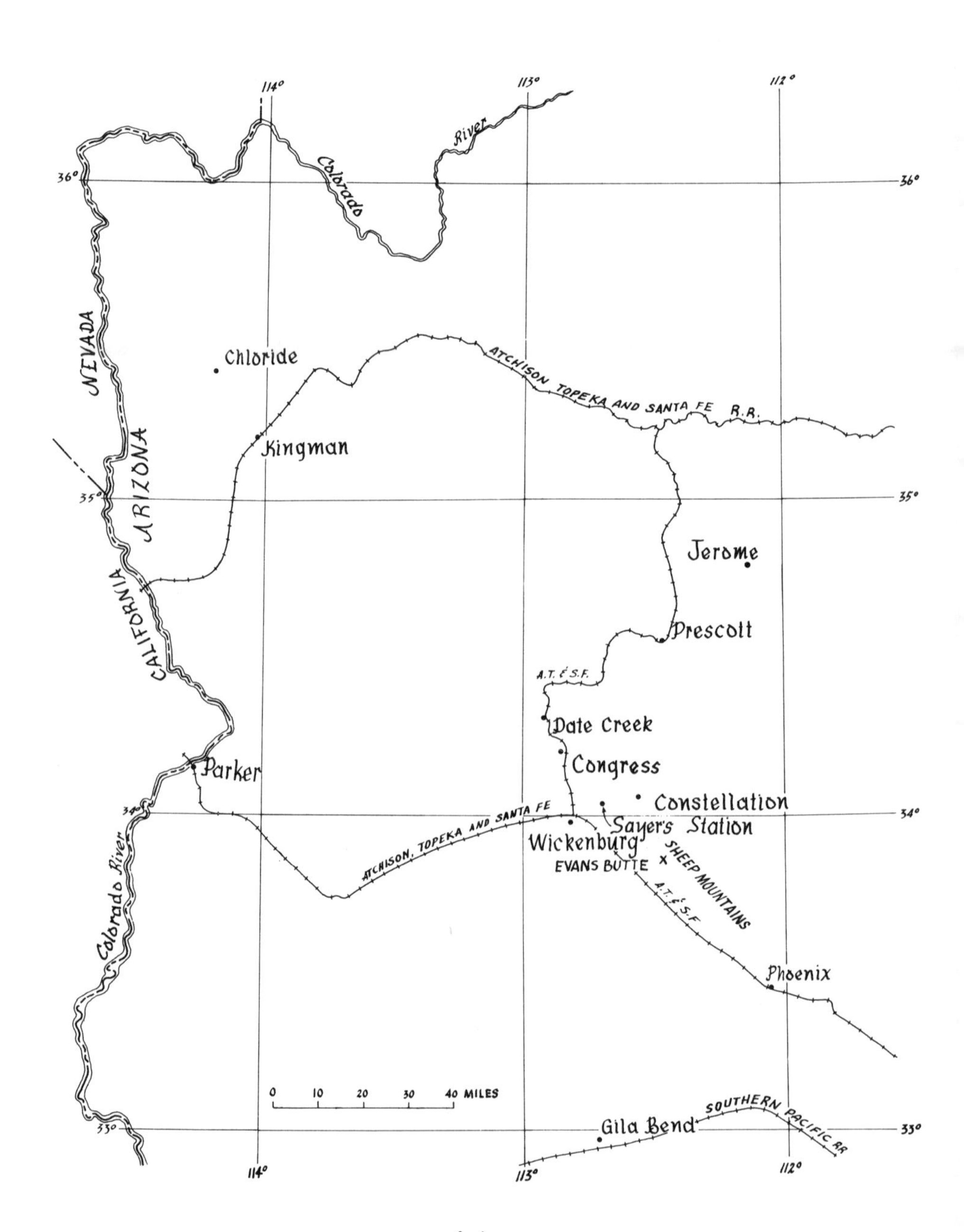
114°
113°
112°
36°
35°
34°
33°
Colorado River
NEVADA
ARIZONA
CALIFORNIA
Chloride
Kingman
ATCHISON TOPEKA AND SANTA FE R.R.
Jerome
Prescott
A.T. & S.F.
Date Creek
Congress
Constellation
Sayers Station
Parker
Colorado River
ATCHISON, TOPEKA AND SANTA FE
Wickenburg
EVANS BUTTE
SHEEP MOUNTAINS
A.T. & S.F.
Phoenix
0 10 20 30 40 MILES
Gila Bend
SOUTHERN PACIFIC RR

Arizona

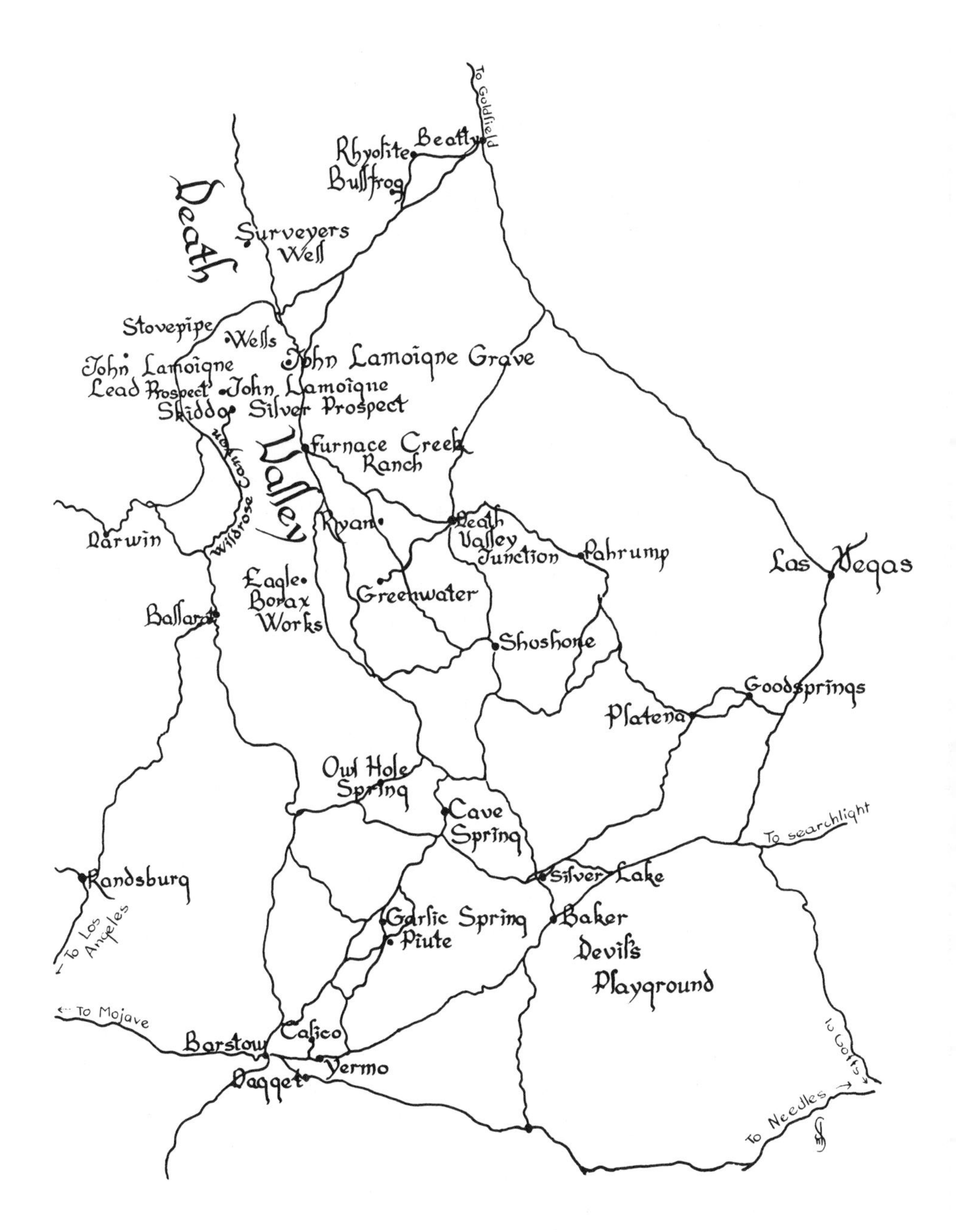

Main Traveled Roads of Death Valley, 1919

INTERLUDE AND A DEAD PROSPECTOR

I was sorry when it came time to leave Estelle's and to surrender the piano to the regular professor, who returned sooner than expected. He had gotten time off for good behavior. It had been a lot of fun, although my fingers tired and were sore from working over the ivories between dark and daylight. The hard-rock and working stiffs who visited the girls were lonely, seeking companionship in the only place they could find it, and the girls were as lonely too. Tragedy and pathos lived apart in a world of make-believe.

The week had been quiet and pleasantly spent. There were few customers, most of them strangers in Kingman with no place to go, other than the saloons, bars, or gambling places, who preferred to keep sober and not throw money over the gambling dump. Estelle's was quiet and congenial. A few of the strangers became regular boarders, and the parlor house resembled a quiet hotel more than anything else. Estelle encouraged the atmosphere resulting from the holiday season, there was no money worthwhile coming in anyway and a rest was more than welcome for her and to her girls.

Estelle gave a welcoming toast to the new year as the strangers and the girls drank to it quietly, and I sat at the piano and played Auld Lang Syne. As she gave the toast Estelle's voice choked and she had all she could do to keep the tears back. A few of the girls left the room, crying, and it was some time before they returned.

Monday following New Year's day, Art and his bride returned from their honeymoon and we went to Chloride together. For all the difference I could notice, work at the mine had been going as well as if I had been there directing it. It was a terrible shock to me. Jack was doing a fine job, as I knew he would; nevertheless my ego was disturbed. As a Christmas present for C. W., Jack had mined and shipped enough ore during the last week of the year to pay expenses for a month.

When I dropped in to wish my blond, and her mother, a happy new year, my reception was something more than frigid. It was colder than Montana during a hard winter. When I passed through the door on my way out, it was slammed as hard as if an elephant had fallen against it. Jack told me later that my turn as a professor in Estelle's was known to everyone in Chloride. The blond's father, who didn't like me even a little bit, had heard about it and retold the story, probably highly embellished, to his wife. And that was that.

The reception accorded me by my blond and her mother was, however, a decidedly isolated event, although my escapade was known to other mothers and their eligible daughters, and when I called to pay a visit, the door was opened before I reached the front porch, and instead of an elephant closing the door as I left, there was a herd of them put to work to prevent my leaving. There was no difficulty in solving such a mystery. Two of the heads of households had been visitors to Estelle's while I was there!

There were, of course, men in Chloride who had not visited while I was substituting for the law-detained professor of Estelle's parlor house, and there was an insistent demand on the part of the elite male population for me to repair to the only parlor house in Chloride and repeat the performance. I was not deterred in any way, and not at all bashful or ashamed of my newly acquired fame. My audience seemed pleased. At least the house professor was shelved for the evening and, in fact, well into the more than small hours of the following morning.

At one point during the evening, when I stopped playing to take five and rest in the small apartment held as a sanctum-sanctorum by the madam for her special guests, my blond's father came to the place, no doubt "accidentally." When I returned to the piano he saw me. There was no way in which he could avoid my seeing him unless he did a magicians' trick and disappeared into thin air. I knew that I had it made. A short and to-the-point talk with him must have been convincing, for on the following day I dropped by again to see my blond and her mother. The welcome of the returned prodigal son had nothing on the one I received. I never learned how the old man changed the atmosphere so suddenly. The formula would have been valuable. Whatever he said or did remains a secret, but my first and last adventure in blackmail was successful.

C. W. received my reports and as a result wrote that he would build a mill and wanted everything in readiness for construction when he returned in late February. A few days after the letter arrived, I pulled out for Prescott to order mill machinery and again left Jack in charge. On arriving in Prescott I wrote John T and Sully to let them know I was nearby. Jerome was less than two hours distant, and they could come down any time they wanted to take five. After mailing the letter, I drifted to one of the Chinese eating places for lunch.

Tom Campbell, whom I had met in Goldfield, was sitting at one of the tables with a man I did not know. When I walked to the table, Tom asked me to sit down. He introduced the stranger as P. W. O'Sullivan, the assistant county district attorney who was to become one of the foremost jurists of the state a few years later.

Tom was getting mixed up in mining in a big way. He had taken a flyer on the Ruby Silver near Caliente in California and was about to reopen the Big Stick, west of Date Creek, as soon as he found someone to manage it for him. The mine had been a profitable one, but operations had been discontinued six years before because the owners wanted bigger cuts than they were getting. O'Sullivan was interested only in the legal side of mining and had recently represented a New York group in their purchase of the Copper Belt gold mine east of Wickenburg. The new owners planned to start operations in May, when the weather was warmer and more settled; O'Sullivan had been commissioned to employ a manager for the coming operations.

That Tom and O'Sullivan were both looking for a mine manager was interesting, and as they talked I did some fast thinking. I had developed a prospect into a mine for C. W. by dumb luck mostly, but I had done it, and then left Jack in charge while I went off on a vacation. Jack had followed my instructions and added something of his own, and the mine went along as if I had never been there. In that moment the mining and metallurgical firm of Crampton and Crampton was conceived. A year later it was born.

When Tom and O'Sullivan were through talking about their problems, I proposed that they employ me to manage both the Big Stick and the Copper Belt. I made it clear, however, that I would not give up my job with C. W. For more than a moment both looked at me as if I were nuts. Then, turning to O'Sullivan and laughing, Tom said: "Well, I'll be damned! Frank did all right in Chloride. I don't know how. But if he has the guts to try the same thing on me I'll take a chance."

O'Sullivan looked at me and said, "If Tom thinks you can do it, I'll take a chance." Then he added, "If you don't make it, you better get out of Arizona."

It took little time to work out the details, and I was the possessor of three jobs instead of one.

Lunch and arrangements with Tom and O'Sullivan over, I sent a telegram to John T and Sully to get the hell out of Jerome and the United Verde and meet me in Prescott. The rest of the afternoon was spent ordering or purchasing equipment for C. W.'s mill, and I took the train to Kingman that night to be in Chloride the following morning. From Kingman I telegraphed my brother, Ted, to get out of New York and hot foot it for Chloride, where Art would teach him how to assay. After this he would go to the Big Stick to do the assaying. Ted had graduated from the military academy and was to enter the ivy-league college to study law, which he didn't want to do. I was going to rescue him from that fate worse than death and make of him a mining engineer and add a few more tons of coal to the family fires burning beneath me. Plans for the mine and mill were gone over with Jack and Art, and I was on my way into Prescott the next morning.

When I reached Prescott, John T and Sully were there. It had taken no time at all to quit the United Verde; the ink was not yet dry on their checks when they met me. Both were put on the Big Stick pay roll and told to round up a crew, purchase equipment and supplies, two ore wagons with trailers, and two six-mule teams. After a couple of days had passed, I realized it would be more than two weeks before John T and Sully had everything ready to go, and that the greater part of the supplies must be shipped by rail. Time was going to hang heavy until time to leave. There was nothing to take me back to Chloride, and John T and Sully had everything under control in Prescott.

Across the main drag from my hotel was Timmerhoff's drug

store. Since the day I arrived, and probably for some time before that, there had been a dust-covered sign in the window: "Boy wanted for soda fountain." I had had no experience as a soda-jerker. In fact, my experience with soda was very limited, but I was certain there would be very little difference between a bar and soda fountain. If there were, I would soon catch on. Timmerhoff was badly in need of help, and he was so busy that he had no time to ask questions, not even if I had seen a fountain or knew about what went on behind one. Within five minutes I had on a white bartender's apron and was at work serving my first customer, for the magnificant stipend of eight dollars a week, fifteen hours a day, seven days a week. It was a good thing that I had enough money to get by on.

Timmerhoff assured me there was not much to the job. I opened the store at seven o'clock, mopped the floor, washed the windows and showcase glass, dusted everything, and prepared the fountain for soda customers. I took the "boy wanted" sign from the window and put it on Timmerhoff's desk, where it could easily be found when I left the job and he would need it again. Timmerhoff arrived at nine and the store was officially open. It was supposed to close at ten every evening but remained open as long as there were customers.

Mornings the store was busy but not crowded—a few customers for drugs and occasional fountain patrons. Timmerhoff was occupied filling prescriptions, and, as a result, most of my morning and afternoon time was spent between waiting on store customers and delivering prescription and drug orders. Some of these deliveries, Timmerhoff thought, took an unusually long time to make; I did not agree with him and thought the time I had been away from the store visiting the customer to whom I had made the delivery all too short. I was kept busy at noontime by a sandwich and fountain crowd, and again from eight until closing. Often I escorted some afraid-of-the-dark lovely to her home after my work was finished.

I learned nothing about the drug business, a little about operating a fountain, but much about customers and their private lives. Almost all of the lovely customers, and a few unlovely males, would ask me to wait on them, or delay making purchases until I was free, no doubt because they did not want Timmerhoff to learn the secrets of their private lives. He probably knew them anyway, but

I was a stranger in town and it mattered much less. Skeletons rattled all over the place and often fell from guarded closets. I did not exert myself to see that the skeletons did not rattle, or that closet doors remained closed. Far be it from me to overlook Sully's advice and the opportunity to learn something, and I had a lot of fun.

The soda-jerker delivery-boy job was well organized by the time John T and Sully announced that everything was ready and pulled out with the teams for the Big Stick. I regretfully resigned my job with Timmerhoff and two days later pulled out by train to meet the wagons at Date Creek, arriving an hour before they did. A dawn-light start was made the next morning in order to reach the mine by mid-afternoon. We had no such luck.

The wagon and trailer loads were heavy and had the full mine crew and their bedrolls on top of the stuff piled in them. It was tough enough going under the best of conditions, but John T and Sully had bought young and badly broken mules, and they introduced us to all of the varieties of hell known to mules with some innovations of their own. Fortunately the two skinners John T and Sully had picked up were among the best of the breed, old Colorado-Wyoming men, who had skinned for United States Army posts where mules were mules and stubborn, and on other jobs where mules knew better. The skinners' conversation in addressing the mules on the way to the Big Stick added something to my vocabulary of profane gems.

We pulled into the mine at daybreak, twelve hours behind our planned schedule. The wagons and trailers were not unloaded, and the hard-rock stiffs who had come out as the mine crew headed for the nearest camp shack with their bedrolls to bunk down and make up the sleep they had lost on the trip out. The skinners were more tired than the working stiffs, but they had no intention of letting the situation rest with victory-obsessed mules. The skinners unharnessed the mule teams and watered them, after which they were close-hobbled, blindfolded, thrown to the ground, and their hobbled fore and hind legs chained together, after which their halters were tied to the hobbled and chained legs as far as their heads could be brought down without breaking their necks. With the preliminaries completed, the two skinners went to work with dray chains.

Each mule was beaten with chains until it screamed. The beatings were administered to each mule several times before the skinners decided it was enough. John T, Sully, and I sitting by watching the performance couldn't understand how the skinners had strength to do it, or the mules to take it. The punishment, cruel as it was, but necessary to prevent some disaster in the future when the mules would refuse to obey, was not over. The teams were harnessed to the still loaded wagons and driven on the roads, up and down hill and at all speeds, until noon, with the skinners rattling chains in the wagonbox every time a mule slowed down or gave any indication of disobedience. The rattled chains were sufficient to bring a mule into the collar and prevent the devil within from getting an upper hand. When it was over, the mules were watered and fed. Not until then did the skinners let up or get any food or rest. After that Tom had the best mule teams in Arizona, the skinners were among the best ever anyway, so there was never again any trouble. The mules were mules no longer; they knew better.

We had gotten little sleep coming out on the wagons, and that night we had less. It was a beautiful moonlight night, but that was all that was beautiful about it. The doors and windows of every building in camp had been high graded. All that remained were the openings in which they once had been, and through the openings the cold wind blew and did its utmost to freeze us to death. But cold wind was not the only enemy of sleep. During the day we had seen a few cats, but, when night fell, hundreds of them made their presence known. The noises were no less than banshees of hell would have made, had they been loose to torment us. It took days to eliminate the feline torment, at two bits a tormentor, which was the bounty paid for each elimination. The hard-rock stiffs did well and accounted for over three hundred. There were more, but I insisted that they be spared in order to keep down the rodent and snake populations.

There was no Sherlock Holmes deduction required to solve "The mystery of the surplus cats." When the mine had been closed down six years before, at least one pair of cats had been left behind. The results were astounding, despite the fact that coyotes had undoubtedly reduced the feline population from time to time. A biologist would have had a field day resolving problems of survival and propagation. There was a spring a mile distant from which cats could get water, but there was not a bird, rodent, lizard, or rattlesnake within miles. Whatever their food or survival problems, they had solved them, and were fine healthy cats with superbly developed voices.

The mine was dry, and six years of desert weather had not worked hardship on the mill and other buildings. It didn't take long to have everything in running order and the mine in operation. In cleaning trash and crushed rock from beneath the mill ore-bin we uncovered four sealed five-gallon tins of zinc-gold precipitate that Ted, whom I had brought down from Chloride, melted to slightly less than $15,000 in gold bars.

Progress on my plan for Crampton and Crampton could not have gone better. Ted had completed his assaying course under Art and was at the Big Stick, Sully was doing a fine job as superintendent, and I was free to go to Chloride and meet C. W. when late February rolled around. On my way to Chloride, I stopped off at Prescott, delivered the gold bars to Tom, and had a talk with O'Sullivan. Tom and I speculated on how the tins came to be under the rubbish and decided that it was probably one of the battling owners who, not being satisfied with his share of the mine profits, had hidden them but for some reason never returned for his loot. However it was, Tom fell heir to more than it cost to put the mine into operation.

Before returning to the Big Stick, I ran down to Wickenburg to take a look at the Copper Belt. I rode to the Copper Belt on the buckboard mail stage driven by Ed Devenney, an old-time driver on the famous stage coach runs of the West. He was putting in his last days on a buckboard stage because the days of Concords were of the past. He had lived his life behind horses and was going to finish it that way. Ed's stage left Wickenburg in the morning and returned in the afternoon, and was rarely without a passenger or two. The Constellation post office twelve miles from Wickenburg was the terminus of the route, with Sayer's Station halfway.

Ed was the best six-gun shot I ever met. On the drive to the Copper Belt we took turns shooting jack rabbits on the run, with six-guns drawn from holster for each shot. Ed did not miss one shot. His targets were brought down with clean upperbody hits. I was shooting as well, or better than I had ever shot before, but there were too many misses. The stage had almost reached the Copper Belt when a big jack made a dive from some brush, headed up a slope, and made the mistake of stopping on the top before going over the ridge. I couldn't have missed that target with my eyes shut, and the head went into the air a foot. Ed kept his six-gun in its holster after that and called it deep enough. Later I learned he passed the word around that I had outshot him. I couldn't have done it on the best day I ever had, or on his worst.

I dropped off at the Copper Belt, and Ed went on to Constellation two miles farther east. There was a watchman at the mine to see that nothing was stolen. Ed was due back in a short time, and I had no time to go underground to take a look at the mine workings. What I saw on the surface did not impress me—the buildings were too large for a small mine, and there was no waste at the shaft dump. There were all of the tools and equipment needed to start work. When John T and I came back to open her up, food, supplies, and powder were all that we would need to get going.

Everything at the Big Stick had gone well, but John T was chafing at inactivity. He had sat around camp the last two weeks watching Sully do all of the work, and he didn't like it. As a result John T and I started out looking for trouble. After canvassing the situation, we decided to look for wild-bee honey. We could, if we were lucky, find some honey, and more trouble than we might want.

There were bees at wet places around camp and near the garbage dump, so we attracted them with wet sugar and broken honey-comb from the mess hall. Whenever bees would sit down on the bait of sugar or honey, they were dusted with flour as the penalty for being too lazy to get honey-making material the hard way, as bees were supposed to do. When a bee detached itself from the bait and started to fly away, a Brunton pocket transit was sighted on her and her course. The job of getting courses was long and tedious, but in several days time, and with patience that would make Job pale with envy, we had several hundred.

When we had sifted the honey from the vinegar, as it were, we had three main courses to follow. We took what appeared to be the easiest, toward a small steep hill with some cliffs that faced the camp on the west which was about a mile distant. It didn't take long to reach the hill, and when we arrived below a low overhanging cliff we saw a narrow crevice through which bees were coming or going, some with sifted flour on them.

After we had looked the situation over, we returned to camp for material for the attack. The assay office furnished all of the chemicals we needed, and the mess hall gave us flour sacks, in which we cut small holes through which we could see. When the bees were defeated, we blasted away the mouth of the crevice and found a cave over two feet wide and thirty feet into the rock. It was fifteen feet from floor to roof. More than three tons of honeycomb were taken from the hive, some of the comb so black that we knew it to be scores of years old. John T, Ted, and I "mined" out all of the comb, gave the stiffs in camp all they wanted, and sent several hundred pounds to Chloride. What was left went to Tom and O'Sullivan.

While we were busy with the bee-hive operation, two prospectors drifted in to camp with their burros. We put them to work packing honey to the cook house. The prospectors were looking for an old timer whom they had met in Congress two years before when he had come into make a shipment of high grade ore to the A. S. and R. smelter in El Paso. When he received payment for the ore, he loaded his string of burros and departed to bring in another load of ore that he said was already mined and ready to sack.

Before the old timer pulled out, he asked the prospectors to help work the prospect with him. His prospect, he had told them, would not be hard to find. It was on a branch of the South Fork of the Santa Maria River and west of the Big Stick. His shack was near a spring, and a few miles almost due west was a table-rock butte. Sully had gotten off of the trail one time when he went out to see how the bee-hive mine was getting along and had found the shack, but no prospector. The floor boards in one corner had been torn up, and what remained of a bedroll lay on a wood frame. A few clothes with nothing in the pockets hung from nails driven in the wall.

Everything in the house was covered by dust, even the burro droppings that the trade-rats had exchanged for pieces of the bedroll and whatever else they had taken. Outside, on the ground and leaning against the house, were a pile of burro saddles, the rope and harness weathered and cracked and the wood of the saddles grey from the desert sun.

When the job of cleaning honey from the hive was over, John T, the two prospectors, and I headed for the butte, to climb it and look the country over to see if we could see anything of the prospect the old timer was working. From the top of the butte we saw the Big Stick and the shack and not far from it three small prospector's workings. They were the only ones we could see, and we were certain that they were what we were looking for. After we climbed down, it did not take long to get to the prospect workings. On one of the dumps were the bleaching bones of a man. They were what remained of the old timer. Buzzards, coyotes, and rodents had cleaned the carcass and scattered bones and remnants of clothing about. The skull had been broken from back to front, probably by a drill steel in the hands of someone who had hit the old timer from behind. His murderers must have known that he had received a considerable sum from his ore shipment and had followed him to the claims and murdered him. The old timer's bones were buried by the sheriff and coroner when they came down to look at what John T and I had found. His identity had been established through his claim location notices.

Since the location had legally expired and the ground was open for relocation, we made out notices in Tom's name and sent them to Prescott to be recorded. The two prospectors who had come looking for the old timer were put to work opening up the prospect and taking out the high-grade ore, which we ran in the Big Stick mill. Because they had been offered a share in the work by the old timer, we gave them a percentage cut of what they would take out during the next year. They deserved something better than a job at straight pay.

Ted (Theo H. M. Crampton), Colorado School of Mines, 1914. Big Stick Mine, Date Creek, Arizona, spring, 1909. Photo by Tom Campbell. Crampton Collection.

Copper Belt Mine Camp, Constellation, Arizona, June, 1909. The shaft, mill, and a few buildings are not shown in the picture. Crampton Collection.

A well-dressed guard (author) in suit and tie, with rifle in hand and pistol in belt, watching for Indians, in Arizona about 1909 or 1910. Crampton Collection.

Wren Mine, Constellation, Arizona, summer, 1909. The author mucking high-grade copper ore after old powder made good. The vein is the dark mass at the right of the Irish buggy. Crampton Collection.

Wren Mine, Constellation, Arizona, summer, 1909. John T, foreground, and the author, single jacking with "dago" single jacks. Crampton Collection.

Wren Mine, Constellation, Arizona. The author getting spring water to pack to the mine, a mile distant. Crampton Collection.

Monte Cristo Mine, Constellation, Arizona, late summer of 1909. Loading high-grade silver ore to haul to Phoenix. From left: Ed Devenney, Joe Bradley, unidentified miner, the author. Crampton Collection.

Wren Mine, Constellation, Arizona, fall, 1909. From left: Alex Trepo, Jim Collins with Irish Buggy, and the author standing by high-grade ore pile. Crampton Collection.

Wren Mine, Constellation, Arizona, spring, 1910. The author climbing from the shaft with geologist's hammer and miner's candlestick with lighted candle. Crampton Collection.

Constellation, Arizona, fall, 1909. Wren's store-saloon in the background. Ah Yat stands beside Ed Devenney's buckboard-stage; between the stage and Wren's store is one of Ed Devenney's freight wagons. Crampton Collection.

At the Cavanass goat ranch, Constellation, Arizona, April or May, 1910. Top row: Dorris with "the" baby, a guest, Mrs. Cavanass, Mr. Cavanass, Mrs. Reeves, Ida Reeves, Mr. Reeves, Bill Durfee. Seated: Dora and Darcas Cavanass. Crampton Collection.

OLD POWDER MAKES GOOD

John T and I went to Wickenburg to meet William J. Dilthey, president of the Arizona Copper Belt Mining Company. Dilthey did not show up as he said he would, but instead there was a telegram saying that he would be delayed until June. The telegram ended by telling me to put on a crew of stiffs to start work. Nothing was said about money. I replied that I would not put anyone on until there were dollars enough to meet payrolls and to purchase supplies, and I suggested that the company establish a credit with the Brayton Commercial Company in Wickenburg. The Brayton Company not only dealt in supplies, but it also acted as a sort of banker for the mines and ranchers nearby.

While John T and I waited for a reply, we spent much of our time in the assay office of Bill Strickland. Bill was English and a darned fine mining engineer, but lung trouble required that he rest in a warm dry climate. Wickenburg seemed to be doing all right by him in health and finances as well. Bill told us something about every mining property in the district. All were in the prospect stage, and he didn't think the promoters who owned them cared very much whether they made a mine or not. Stock selling was their business, not mining.

The Copper Belt, Bill told us, had been purchased from Denver promoters with whom Dilthey had at one time been associated. Dilthey was a New York architect and a good one; but he thought gold came from the ground in solid chunks or bars, and Bill cautioned us not to expect much from him. He did say that the Copper Belt and Monte Cristo were the best looking prospects in the district, and he thought they had the makings of a mine.

In Dilthey's reply to my telegram, he instructed me to get the mill going and to start production imediately. He said that money was being sent to the Brayton Company, enough to pay expenses until payment was received from the sale of mill concentrates. It was evident that Dilthey was inexperienced in the cost of a mining operation. John T and I gave up and decided to do nothing until Dilthey arrived in June.

John T and I went out on the stage with Ed Devenney, dropped our bedrolls at the Copper Belt, and went on to Constellation. P. S. Wren, "P" for Powhattan, who owned and ran the Constellation establishments, was a five-foot-three edition of a Kentucky Colonel; he was the embodiment of everything a Southern Gentleman was pictured to be, including a well-groomed mustache and goatee, his especial pride. He wore high leather boots, black well-creased trousers, and a cutaway morning-coat; under the coat was a cartridge belt with a holster that held a Colt frontier forty-four. I never saw Wren use the six-gun on a live target, but on tin cans and bottles thrown into the air he was almost as good as Ed Devenney.

Constellation boasted several buildings. The largest—a two-story, multi-room affair—had been built as a stage station when Constellation was the center of an active and busy mining camp several years earlier, but Wren had put it in use for purposes far from those for which it was intended. The dining room was still used for serving meals, but the office and lounge had been converted into a dance hall. A tinny piano stood against one wall, and a professor somehow coaxed tunes from it for dancing. Each of the station's rooms was occupied by a girl from the Phoenix stockade, who had come for a few weeks rest and to do a little business on the side.

The stage station rest home was up a dry wash, a hundred yards from the gulch in which Wren's store-post office-saloon was located. The post office was a two-by-four cubicle walled off from the bar-saloon part, with the bar on one side facing the store's shelves filled with food supplies. Between were a couple of poker tables, a twenty-one table, and a faro bank. Behind the building was another, a three-room affair, in which Wren and his assistant bartender-clerk held forth during sleeping and eating hours.

Up the gulch were a small corral and shed where Ed fed and watered his mail-stage teams, and a little farther up the gulch were a larger corral and sheds in which Wren kept a string of saddle ponies for his resting girls to ride. There were almost always other than Wren's ponies in the corral over night. Almost every operating prospect had a pony or two, and the stiffs who worked at them took turns riding over to see that the girls were getting the rest they should.

Wren had an around-the-clock business to which he attended long hours at a time with a helper, Ah Yat, a Chinese of more than ordinary accomplishments. Twenty years later I met Ah Yat again, in China, in unexpected and unusual surroundings. Ah Yat was the Master of the Monastery at Hon Lock, a few miles down the Pearl River from Canton, on Ho Nan Island. During the many years I lived in China, I visited Ah Yat often.

Wren's saloon and stage station were busy most of the time, but especially so when the stiffs from nearby mines came in after shift and on pay days. There was nothing for the hard-rock stiffs in Wickenburg—the only settlement worth while nearer than Phoenix—that was not available to them at Wren's, and they took advantage of it.

Ah Yat worked behind the bar and in the store from early morning until late in the evening, and called it deep enough only when he could not take it longer and had to hit the hay. Sometimes Ah Yat spelled Wren in running the faro-bank or twenty-one game, but most of the time the stiffs played stud with the house taking the cut that Wren demanded, trusting one of the stiffs or a girl from the rest home to collect for him, but Wren kept a weather eye on the players to keep them honest in taking out house cuts.

Wren did not suffer from want of help, and it was not often that he was required to perform chores around the store. He did, however, attend to his postmaster duties meticulously himself. His bar duties were mostly in serving drinks to himself. When either he or Ah Yat had to call it deep enough and hit the hay, one or more of the girls from the Phoenix stockade served drinks to the stiffs, or worked at the tables, seductively and efficiently. When one of these girls departed for the rest home quarters with one of the stiffs, there was always another to take her place.

Wren was full of whiskey and information. While we waited for the stage to pull out and take us back to the Copper Belt, he became loquacious. Before we left, he had unburdened everything there was to tell about in the district. Wren was all for the prospect-mines, and hard-rock stiffs, of whom there were several hundred working nearby. Wren's rest home was doing a land-office business, and it didn't appear to me that the girls were getting the rest they had come for.

Cattle men and goat ranchers were Wren's pet hate. The reason for his gripe came out little by little while he told us about them. There was but one cattle ranch, which ran only a few hundred head of steers and was not worth mentioning, although it had four cowhands that he didn't like. He liked the owners of the three goat ranches less, with their score of herders, the women who were hired to cook for them, and the rancher's families. The largest goat herd was owned by a rancher who had three daughters, only one of whom was married; the unmarried two, Wren told us, should get married. He did not care for the other goat ranchers, either, but he was less emphatic than when he talked about the larger of the three—the owners of the smaller herds had only one daughter each.

None of the cowhands or goat herders came to Constellation for anything but mail, and never remained longer than was required to purchase supplies and to open and read the letters they received. Not one of the cowhands or goat herders ever bought a drink, nor did they take advantage of the rest that Wren's station girls offered.

Wren's opinion of cattle and goat ranchers seemed to be shared by most of the mining stiffs and all of the rest home girls. The mining stiffs had more than a reasonable gripe. The herds of goats would mill past the holes they were working in, and loosened rocks would fall into them and sometimes cut or bruise the men badly. The rest home girls had a gripe, too; the cowhands and goat herders did not pay them the attention they should—they were not at all sociable.

Wren was in a salesman's mood. He had two mining claims not far from Constellation, and he offered to let us have them for ten thousand dollars. We felt complimented that Wren would think we had that much loose money, but made the excuse that we didn't want any more claims than we already had. We did not tell Wren that we owned none. Ah Yat, who had overheard the conversation, came over when Wren stepped away for a few moments, and told us that one of the rest home girls had gotten the claims from an old prospector on a binge, and had sold them to Wren for fifty dollars. When Wren returned, we offered him one hundred, not expecting him to accept; but he did, as Ah Yat looked over and beamed.

Two of the girls of the "rest home" rode up minutes after we became mine owners, dismounted, and came inside. Wren looked at them, and then at the unoccupied men standing at the bar, but said nothing. His look was all that was needed. The girls walked to the bar, and when the men there made them welcome, Wren again became his old self. He was money hungry and didn't want to lose sales at the bar, or at his stage station for special service to temporary occupants. Then, probably thinking of dollars he had not made while the girls were out, he offered us the ponies, with saddles, that they had been riding, at a price no one could refuse. We cut the offer in half, and, when the stage pulled out, it was loaded with hay and oats to be dropped off at the Copper Belt. John T and I rode back on the ponies. The animal I picked was a red pinto. I called him Red. As we left, Ah Yat stood in the doorway, beaming from ear to ear; there was no question that he approved of our bargaining methods.

The next day John T and I looked over the Copper Belt and promoted the watchman to camp cook, which job he did well, as any hard-rock stiff or desert rat would. The tent houses and frame buildings looked as if they belonged to a large operating mine. There were tent houses for the men, individual quarters for the bosses, and a mess hall fitted with everything. The office building was huge and contained fully equipped assaying and engineering offices. Every tool imaginable was in either the blacksmith shop or tool shed.

The head house and gallows frame were on the highest point of the vein, with ore bins and a mill building below it. The "mill" consisted of two batteries of one thousand pound stamps and copper plates; there was nothing else. The mill looked fine in a picture of its exterior, but it would take twenty thousand dollars for more machinery before ore could be run in it.

The shaft was an incline, to which a small trapdoor permitted access to the shaft ladderway. A prospector's windlass, wound with rope, at the end of which was a small ore bucket, was mounted on the shaft collar. One plank had been removed from the shaft cover and the shaft used as the camp privy. A lot of men had been employed, at one time or another, but there was no evidence of any work having been performed underground, excepting that which was related to the privy.

On the hillside, below the shaft, a wide gold-quartz vein outcropped for fifteen hundred feet. If samples had been taken of ore in the outcrop or shaft, there was no evidence of it; certainly there were no assay records or maps. Since a lot of work was to be done, I wrote Ted to come down from the Big Stick. I sent the assistant assayer from Strickland's office in Wickenburg to take his place, and Bill was pleased to have him get a better job.

John T and I did not like the looks of what we had seen at the Copper Belt. What we found followed the same pattern as that of hundreds of other good prospects. It was an old story: promoters had mined the sucker-investors instead of ore. I knew that Dilthey and the new owners knew nothing about mining, and had I not thought the Copper Belt could have been developed into a mine, John T and I would have called it deep enough and told Dilthey to go to hell. Later I wished we had not waited.

I wrote O'Sullivan and told him what we had learned and had the nerve to tell him to lay off having anything more to do with the property until Dilthey gave signs of doing something substantial. O'Sullivan thanked me and said that he would not have had anything to do with either the old or the new owners had he realized what had been going on.

Despite the shaft's having been used as a privy, John T and I went down. It was one hell of a way to put in a first shift, but the job had to be done, pleasant or not, privy or no privy. At fifty feet the shaft passed through the vein where it dipped into the foot wall, and at one hundred, where it was bottomed, drifts and cross cuts had been run in every direction in a futile effort to locate the ore again. The effort indicated that whoever was doing the work didn't care whether the ore was found again, or possibly didn't know where he was going.

It didn't take long to look over the work on the hundred. There was not much to look at. If we had any question in our minds before we looked the thing over that the former promoter-owners had done no work underground, the uncertainty was wiped out when we came to the end of the drift. In the face were a few prospector tools, a battered-up Irish buggy, a half-empty box of blasting caps, and a couple of rolls of fuse. On the drift floor, beside the Irish buggy, were three cases of forty per-cent powder. One case had been opened and a few sticks taken out. The other cases were unopened.

John T and I looked at the powder in the opened case and backed away; neither had to urge the other to get going. The powder had been there a long time, and the time could be measured not in months but by years. Obviously neither the Denver promoters nor the Dilthey crowd had done any underground work, and had probably not even been down the shaft. I did not blame anyone for not going down through a well-filled privy to get to the bottom of the shaft to see what was there. The powder and tools were those of a prospector, possibly the original locator-owner who had sold the claims to the Denver promoters years before.

The sticks of powder in the opened case had sweated through their wrappers on which we saw myriads of shining crystals of nitroglycerine. Those three cases of powder had waited in patient resentment at not being used, until someone should come along and move one of the sticks, or give the cases a jar. When that would happen that joyous one hundred and fifty pounds of violence, sudden death, and destruction, would let go and live up to its reputation. No one close to it at the time would again care whether its reputation was good or bad.

John T and I wanted as little as possible to do with our discovery, but it must be gotten out of the shaft, and it was our job to do it. For the time being, it was safest where it was. John T and I were in no hurry whatever to move that powder; in this case last things should be done first, and possibly first things not at all. The privy was bad enough, but to find this murder in the shaft was too much. To get the stuff out we must go through the privy three times, each way, and risk our lives doing it. We were not going to be hurried into going back into the double dose of trouble.

For two weeks after we found the powder, John T and I rode our ponies all over the country and visited every so-called mine and prospect hole we could find. In doing the rounds we took a look at the claims we had purchased from Wren, and from what we saw we concluded that we had lost one hundred dollars. There was not a prospect hole on the claims, but the monuments holding the location notices could be seen for miles. An outcrop, a few feet wide, had blue and green copper carbonate stains, and not much of these. Of one thing, however, we were positive: the vein, or whatever it was, had not been salted.

About the time John T and I had looked at all of the holes and prospects within miles, there came a telegram from Dilthey. He was again delayed and would not arrive until July. It had been a month since I had been at the Big Stick or at Chloride, so I went to see how two real mines were getting along. The Big Stick was going fine, the mill in Chloride was going up fast, and C. W. had returned to New York. Jack had pre-empted my blond, and there was no reason to stick around; so I headed back for the Copper Belt to wait another month for Dilthey to arrive.

Dilthey had telegraphed more money to the Brayton company, but it was not enough to do anything with; so John T and I had more time to kill. While we waited for Dilthey, we wanted something to do to keep us out of the Copper Belt shaft-privy and away from the old powder. We were looking for good reasons to delay the unfortunate day that we must go to work on the old powder, although poor ones would do as well. We put our problem up to Ah Yat one evening, and he suggested visiting the cattle and goat ranches. The idea did not seem to be a good one, but we would have something to do to delay the shaft-privy job.

The lone cattle ranch, owned by F. X. O'Brien, did not interest us. We found no excuse whatever for visiting it again. With the goat ranches it was a different matter. After visiting them we thoroughly disagreed with Wren's opinion of goats and goat ranchers. We saw nothing at the goat ranches that could possibly be termed a nuisance, even the goats that might cause rocks to fall into shafts. At the goat ranches there were all of the reasons we needed to delay our adventure into the Copper Belt shaft. In fact, there were four of them.

The owner of the largest and finest goat herd had two attractive unmarried daughters. The married daughter was expecting an arrival before the summer was over. The daughters of the other goat ranch owners had no less attractions than those exhibited at the larger ranch. The goats, however, didn't measure up to the same standard as at the larger ranch, which was a decidedly unimportant detail.

After making our discoveries at the goat ranches, we filed location notices and did some prospecting. Ah Yat tried his Celestial best to learn why we were visiting Constellation less often. Ah Yat appeared satisfied when we told him we were busy and a two-mile ride from camp was too much after a shift was in. We failed to tell him that the goat ranch where most of our shifts were put in was six miles from camp. Ah Yat knew we were lying, but couldn't figure where John T and I were spending our time, although he knew it was not at the Copper Belt.

After we had spent three weeks of our time at the goat ranch, we reluctantly gave some thought to the old powder in the Copper Belt shaft. It took time to get the one hundred and fifty pounds of sudden death up the shaft ladderway and to the surface. We first poured water on the powder in the opened case in an attempt to dissolve the nitroglycerine crystals, then turned to the two unopened cases and gave the powder in them the same treatment. Taking the tops from the cases would have shamed a mother who was handling her first new-born baby for the first time; the care that the mother gave her baby would have appeared as mayhem in comparison with the tenderness we exercised.

With the precaution of men walking on broken glass, without shoes, to prevent having their feet cut to ribbons, we carried the cases of sudden death up the ladderway and out of the shaft-privy. An improvised stretcher received our uncertain and deadly loads. Several quilts were placed between the stretcher shafts for the cases to rest upon, over them other quilts, and around, over and under, we lashed a canvas tarp. Regardless of whether the nitroglycerine crystals were potent or not, we were taking as few chances as possible of leaving the Copper Belt and Constellation as sausage meat.

A few days before we brought the powder out of the privy, John T and I dug a trench in which to place the stretcher, on one of the copper-stained outcrops at the property we had bought from Wren. It was our intention, whether the claims were good or not, to complete location and development work on the claims at one fell swoop.

Before John T and I left the Copper Belt with the old powder. I put a ball of cotton waste at one end of the stretcher, and covered it; at the other end, a pair of worn-out shoes, uncovered. John T asked why I did such a fool thing, and I told him it would remind us to be careful, since our load now had such an excellent resemblance to a dead man being carried on a stretcher.

When Constellation was reached, we deposited our load outside of Wren's and went inside for a drink. We needed one. Soon some of the stiffs went out to put in some more profitable time at the stage station, and with them some of the girls. They were back in a matter of seconds, asking who the dead stiff was, and a lot of questions we couldn't answer. When the thing had gone far enough, we told them what was on the stretcher, and what we were going to do with it. When we left Constellation there were not a few who went with us to see the show. Ah Yat was in the caravan, at least a dozen hard-rock stiffs, and a few of Wren's rest home girls.

John T and I wouldn't let anyone help carry the stretcher. We were taking no chances that someone with an over-quenched thirst would stumble and the powder would go off to send the procession to hell and other places along the way. There was no one who insisted, however, and the stretcher followers were usually several hundreds of yards to the rear. The girls were slow walkers, and there was not a stiff who did not show his gallantry by remaining with them to assist them over rough places along the way.

When the claims were reached, we laid the stretcher in the trench on the outcrop and prepared to shoot. Meanwhile the crowd that had followed climbed a hill about a quarter of a mile from where the blast would take place, and from which an excellent view of proceedings could be had, with little chance of flying rocks reaching them. John T cut holes in the tarp and quilts and eased new powder and primers between the cases on the stretcher. John T spit the fuse.

The fuses were cut to burn fifteen minutes and we had plenty of time to get away and were safely behind a hard-rock outcrop half way up the hill when the thing went, two hundred feet into the air, and clouds of dust in every direction. As soon as rocks stopped falling and before the smoke and dust cleared, John T and I were on our way to the hole the old powder had dug. It was a beauty. A couple of elephants would have been lost in it. Meanwhile the stiffs and girls were on their way from the hill, led by Ah Yat.

The blast opened a three foot vein of rich copper ore, the top of which was formed as if it were a huge mushroom. On his way from the hill, on a run, Ah Yat had picked up a piece of the ore that had been thrown out by the blast and was prepared, not only for what he expected to see, but also for us as well; he arrived at the hole minutes before the others.

Ah Yat did not hesitate in opening the subject and was not interested in looking at the ore the blast had opened up. Already he was aware that good ore had been found, for the piece he had picked up told him all he needed to know. Ah Yat wanted to be cut in on the claims. He not only wanted it, but declared himself in, without more than a superficial effort at politeness. We gave him no arguments, for he had tipped us to what Wren had paid for the claims; and he had been responsible for our purchasing something we did not want, and hadn't expected to get, when we cut Wren's offer from ten thousand to one hundred dollars.

The loss of the one hundred dollars would not have hurt us had the claims proved worthless. A month before, when we took our first look at them, two bits was all we thought they were worth, which, we thought at the time, would have been a robbery. So we cut Ah Yat in, for he certainly deserved thanks. Ah Yat did all right for himself that day, and we did not do so badly ourselves. A few days later the Wickenburg *Miner,* the town's only newspaper, ran a story about the blast; its headline: "Old Powder Makes Good."

MIDWIFE AND GODFATHER

A few days after John T and I disposed of the old powder, Ezra Thayer, owner of the Monte Cristo silver mine, which adjoined the Copper Belt and had its camp less than a mile away, drifted in from Phoenix and came over in time to be invited to supper. It was not long before we knew that Thayer had something on his mind, but it was not until we were through eating that he opened up and got it out of his system.

Years before Thayer took it over, the Monte Cristo had been worked by Mexicans who recovered large amounts of silver from ore they mined and smelted. The kiln in which the ore was treated was a short distance down the gulch from their workings; around it were tens of tons of slag, mute evidence of the large quantities of high-grade ore that they had taken from their prospect holes and smelted.

While the Mexicans were making a small fortune for themselves by smelting hundreds of pounds of high grade every day, a couple of prospectors learned that they were not American citizens and therefore had no legal rights of ownership in the claims. The prospectors had no regard for moral rights and exercised their legal ones by locating the claims, after which they drove the Mexicans off after a few days of gun-play but no shooting.

The prospectors got what was coming to them, for after prospecting the claims for months, they found no signs of the area where any of it had come from. Apparently the prospectors owed Thayer money which they couldn't pay, but, owe it or not, he paid them something, took the claims over, and the former owners departed.

Thayer had no more luck than his predecessors, excepting that in the shaft he sank there were very small lenses of high-grade ore in the low-grade vein and sometimes a larger slab on the hanging wall. Thayer was certain that there was a high-grade vein somewhere, but he was also having a hell of a time convincing himself that it would be found without spending a fortune on prospecting and development work.

On the side of calling it deep enough were his hard-rock stiffs, who were Cousin Jacks. If there was high grade within miles, their instinct would have taken them to it, but they had found not a sign of the kind of ore the Mexicans had been taking out. On the side of his continuing to keep looking for the rich ore was the knowledge that the Mexicans had been taking it out and smelting it up to a few days before they were kicked out. Thayer did not want to be called a fool for continuing to spend money and ending up by finding nothing. On the other hand, if he did call it deep enough and pulled out, he didn't want someone else to come along and find the high grade.

Thayer wanted John T and me to take over the job of finding out where the high grade the Mexicans were taking out came from. If we couldn't find it, no one could. Thayer had a great deal more confidence in us than we did in ourselves and told us the reason. John T and I had found high-grade ore where no one had any idea it existed. It was impossible to convince Thayer that his opinion of us was hogwash. He refused to believe us when we told him that the only reason we blasted the powder where we did was only to get the claim location work done and the assessment, or annual requirement of the government, work over with. When our arguments failed, even when we exaggerated the truth, there was nothing to be done but accept Thayer's proposal and hope for the best.

We were up at dawn and off to the Monte Cristo claims. We made a wide detour around the camp so that the sound of our ponies hoofs would not be heard. We had been over the claims and almost all of the outcrops on foot, never on a pony, and everything looked different from the height of our saddles. We had not ridden a hundred yards along the outcrop that the Mexicans appeared to have worked on the most, when we discovered what we were looking for.

The Mexicans had mined the high-grade ore seam that lay two

feet above the three to four foot wide low-grade vein. There was barren country rock between the larger and the smaller veins. The Mexicans had skillfully hidden the high grade seam by covering it with two-by-twelves and then throwing the worthless low grade from the vein over them. We removed one of the two-by-twelves and crawled into a narrow trench about thirty feet long. At both ends of the trench and in the bottom was the high-grade ore, in a seam from one-half inch to sometimes over two inches. It was a beauty, and we helped ourselves to a pocketful, crawled out, replaced the planks, and re-covered them with waste. Then we rode back to the Copper Belt, before any of the stiffs there, or at the Monte Cristo, were awake to see us.

After breakfast we rode over to the Monte Cristo, this time openly and with as much noise as we could. The rest of the morning was spent riding over the claims, as far from the place where we had discovered the ore as we could. From time to time we would dismount and hammer at barren rock with our prospecting picks, then sit down to talk, gesturing and pointing in every direction as we did so. We hoped that Thayer was watching us and would think we were really doing more than killing time. We had no intention of letting Thayer know for a week or ten days that we had found what he was looking for within an hour after we started out to find it for him.

John T and I spent most of our time at the goat ranches visiting with our discoveries there, but we did spend a little time in the mornings and sometimes in the afternoons wandering over the Monte Cristo claims where Thayer was sure to see us. When we had killed as much time as we thought would be necessary to convince Thayer we had really done a job, we told him where to look, in the shaft eighteen inches to two feet into the roof. We did not disclose that we had found the place the Mexicans were mining the ore that they smelted, but left that for disclosure to Thayer at a later time when we would get certainly more credit for being "mining experts."

Supper was over when we told Thayer where to look for his ore, but there was no waiting until morning; he wanted to determine whether we were right or wrong, or only guessing. The hard-rock stiffs were called out, sent into his shaft to drill short holes and, when that was done, to shoot them. When the powder smoke from the shots had cleared from the shaft, we went down. Some of the ladders were broken, skids for the incline-shaft bucket bashed up, and the hole an all around first-class mess with high-grade pieces, some as large as a corrugated wash-board, scattered through the mass of gob and low-grade ore from the main vein. Hanging from the roof were slabs, some as big as a door, of horn-silver, one of the richest of silver-ore minerals. One was removed later and used as a soft-toned bell.

It took two weeks to take the high-grade from the roof of the shaft and at the same time timber to keep it from caving. When all the high-grade had been taken out, Thayer loaded it into two ore wagons to be hauled to Phoenix. Ed went with the lead wagon and John T in the other. Both were good six-gun men, and Thayer took no chances of theft of his silver ore. Thayer and his hard-rock stiffs went along on ponies as supplemental guardians, while I remained behind to be handy in the unlikely event that Dilthey showed up.

A few days after John T had left, Bill Durfee, one of the goat herders, drifted into camp, blood streaming from his left arm and his clothes a mess. When Bill had reached to pick up a stone to put in the sling he used for herding his goats, a rattler that he had not seen had struck and got him in the left arm above his wrist. Bill killed the rattler, put on a tourniquet, cut three deep gashes across his arm through the snake fang punctures and then went to work sucking the poison out. After Bill thought he had gotten all of the poison he could, he still was not satisfied and took a block of sulphur matches, and when the sulphur was afire plunged the burning mass into the snake and knife wound to cauterize it.

Bill had had to walk less than a half mile to the Copper Belt but was in bad shape when he reached my office door. Ted and I poured raw whiskey into Bill, and before it was all gone, Ted saddled Red and rode to Constellation for more. When Ted returned, we sewed up Bill's wide open wounds with boiled string and a curved ore sack needle, put heavy bandages around the arm and kept them soaked with hot water until Bill showed signs of becoming better. We put him into the saddle on Red and took him back to the goat ranch. Ted and I walked beside him to keep him from falling off. What was left in the second bottle of whiskey we took along for emergencies.

The next morning I saddled Red and rode to the ranch to see

how Bill was making out. As I rode down the gulch toward the ranch, I noticed that the goats were in the corrals and that both the ranch buckboard and wagon were missing, as were the saddle and wagon animals. I guessed that Bill must have taken a bad turn and everyone had gone with him to Wickenburg. I rode up to the ranch water trough to let Red have a drink and was about to pull out when I heard moans coming from the ranchhouse.

I halloed and got no answer. When the moans continued, I went off Red and into the house, calling as I went down the long hall until I heard moans from a room near the far end. On the bed, tears streaming down her face, was Dorris; she managed somehow, between moans, groans, and writhings, to tell me that her long-expected baby was about to arrive and that everybody, including her husband, had gone to Wickenburg to take Bill to a doctor and attend a dance following some sort of celebration that was going on in town.

For a moment I cursed myself for inviting the ivy-league college to bust me because I didn't want to become a doctor. Nothing could be done about that now, and it was up to me to face it alone and unaided. Dorris was about to have her first baby, and she knew no more about what to expect than I did. Although her woman's instinct told her something about it, that instinct wasn't worth a damn because she was too full of pains for it to go into action.

Between moans and with almost unintelligible words from Dorris, I did learn where the things for the expected baby were cached. I got them out as quickly as I could, after first washing my hands in carbolic acid diluted with water. My state of mind was clearly brought to my attention when I realized the acid-water dilution was something more than a chemical error, and I had to go through with the job with painfully burned hands.

The women of the family had planned to deliver the baby without a doctor; I was certain that if they could do it, so could I, but I was scared stiff. The only thing I didn't find in the cached stuff for the baby's arrival was borax, which I understood was used in a solution to wash out its eyes immediately after birth. I remembered having seen specimens of disintegrating borax ore on the ranchhouse porch, picked up probably by one of the herders. I took one of the easily crumbled specimens from the porch, put it in a pot on the stove and let it boil, beside the wash-tub of water that I had put on earlier.

Before I had everything arranged where it could be reached quickly on the freshly sterilized table, Dorris let out a yell. I got over to the bed in a hurry only to find that it was not the baby arriving but simply that she had had an unusually severe pain. When I returned to my laying-out and arranging job on the table, I realized that I needed more minutes to get my courage and nerves restored to something near to working order.

My worst nervous guess was nothing as compared to what was actually to overtake my stumbled-on new profession as a midwife. Dorris was giving signs that something was about to happen, and her pains were making her writhe and moan more than I thought she should. I remembered when I was a boy and had stomach pains that if I braced my feet the pains let up. I eased Dorris to the foot of the bed, against which she pressed. The pain started to ease, and I turned to complete my chore of arranging things on the table. But time had run out on me. Dorris gave forth an unusually long moan followed by an extended "Oh-o-o-h, Oh-o-o-o-h," and I turned to see what was going on. The baby was on its way without the help of Dorris or me. I was beside Dorris giving the baby as much help as it needed, which fortunately was almost none at all.

When finally Dorris completed her first job of motherhood, I was leaning over holding a squirming but silent ball of deep red flesh. I knew that the cord must be cut and tied, and although I had forgotten it up to that moment realized something must be done about it in a hurry. I didn't know whether to make the baby cry or to cut the cord first; so I compromised by turning the baby upside down and spanked it, and with the cry that followed came liquid from the baby's mouth, and I was sure I had done the wrong thing first. There was no time to think about that. The baby was still crying; so I reached for the scissors on the table, after tying two threads around the cord, and cut it between them.

Nothing happened. The baby kept on yelling, and I knew I did not have a murder on my hands. Then I laid the baby down, washed its eyes with my pseudo-borax solution, poured out some olive oil, washed the baby with it, and wrapped the new born in a piece of flannel from the junk on the table. I thought my job was finished, and I turned the baby over to Dorris.

I was taking five and congratulating myself that I had had sense enough not to be a doctor, when all hell broke loose. If I had had a case of nerves before, they were nothing compared to what came next. I was certain that although I had not murdered the baby, I was going to have a dead mother on my hands.

Nothing I had studied in mining or anything that I had heard about prepared me for what came next, nor did I know that the birth of a baby was not the end. Somehow the thing was managed with Dorris holding her baby and looking as if she were a Madonna while I sweated and hoped that she would live. Things finally cleaned up, my nerves returned to nearly normal when I realized that Dorris and the baby were both alive. There was comfort in the knowledge that I had not committed murder, but I was certain I had done something wrong and that Dorris was in a critical condition.

Many years later a doctor explained that everything that happened had been going on for centuries and was normal, and it was also encouraging to know that my only mistake was having Dorris brace herself against the foot of the bed and push. The pushing business was a mistake in the right direction: the baby had arrived before Dorris and I were prepared, and it was much easier for both of us because of it.

The baby took turns at yelling and sleeping the rest of the morning, through the afternoon, and into the night, while Dorris cared for it and I cared for Dorris. The family and ranch hands did not show up until after midnight, more tired than I, and surprised that the baby had arrived. There had been no sign of anything like that happening when they departed early in the morning. I was glad to turn my midwife duties over to the family. I had gone through enough and was glad to pass the buck to someone else.

With the buck passed, I asked about Bill and learned that he was getting along all right but it would be some time before he was fit to herd goats again. I had completely forgotten about Red, but he had drifted to the water trough and had not wandered. I put him in the corral, unsaddled, took the saddle blanket, and turned in on the ground at the side of the ranchhouse. I had had enough for one day and needed some sleep. The ground by the ranch house suited me fine.

After breakfast I was ushered into the birthroom to see Dorris and the baby. It had weighed six and one-half pounds that morning. It was not until the family was gathered in the room to notify me that I was the godfather of the baby girl, that I had known the the sex of the baby. I had been too occupied with Dorris to notice, and Dorris too happy and miserable to ask me.

John T returned from Phoenix and came in on his pony the day after my session at the goat ranch. Thayer, John T told me, had placed two mine ore cars in the window of his hardware store and filled them with Monte Cristo high grade. Between the ore cars was an Irish buggy filled with Wren copper ore. The Monte Cristo ore that Thayer could not get into the window made an impressive pile in the back of his store.

News of Thayer's strike at the Monte Cristo got around in a hurry, and word of it spread to the east coast, as did discovery on the Wren. It was not long before promoters drifted in to make themselves nuisances and to clutter up the landscape looking for high grade outcrops and veins. The number of tenderfeet became so great that it was almost impossible to go anywhere without stepping on one of them.

Ed Devenney was in the height of his traditional glory and took a new lease on life, sometimes making two round trips a day between Wickenburg and Constellation. The tourist-promoter traffic had become so heavy that the buckboard couldn't handle it all in one trip. Wren complained that he was overworked and that the tenderfeet were a damned nuisance, but he was nevertheless pleased that large chunks of "foreigner's" money were coming into his establishment; and instead of one six-gun he wore two as he drifted among his gambling tables acting as lookout for them all. Not only did Wren have a twenty-four hour shift operating the gambling tables, but he also had to find a full-time helper for Ah Yat, because business in the rest home for girls from the line became so brisk that none had any spare time to help behind the bar or at the game tables. He had to send a rush order to the Phoenix stockade for others to come up to join the "rest home."

Unlike Ed and Wren, who were thoroughly and profitably enjoying themselves, John T and I were having a hard time avoiding the nuisances who had drifted in from the East to see what they could pick up as sucker-bait to bolster stock selling campaigns on companies formed to develop worthless prospects. We had no time for the promoter breed and refused many offers to use our reputed talents in finding high grade for them. It was easy to refuse but a hell of a job to keep them from bothering us. There was no longer any place we could hide. Even the goat ranches were cluttered up with tenderfeet and having troubles of their own keeping the promoters from digging holes all over their ranch lands—and from shooting goats because they thought they were wild.

In desperation John T and I took to the hills immediately after unusually early breakfasts until late at night and rode miles hunting coyotes and foxes, or anything else that we could find. One night, when the moon was bright, John T and I decided to test the Indian legend about the Hassayampa River and drink from its waters on bended knees in the middle of the stream, facing its source, at midnight when the moon was full. We became soaking wet during the drinking ritual because we were meticulous about every detail; from that moment we must wait to see whether, as the legend ran, we would always be wanderers, never have friends, nor possess wordly goods, and have misfortunes follow us wherever we went. The legend may be right or wrong—there are a few years left to find out—but at least the wandering part has been fulfilled, and it has been a lot of fun.

Another night with the moon so bright we could clearly see objects at several hundred yards, a pair of coyotes jumped a jack in a wash below where we were riding, and we pulled up our ponies to watch. One of the coyotes took after the jack while the other ran wide to head it off. Whenever the jack turned sharply, the coyotes reversed positions in the chasing and heading-off tactic. The maneouver of reversing the action, chasing and heading-off, continued for some minutes until the circle in which the jack could move became so small that its last desperate effort to run free ended in the jaws of a coyote. What had heretofore been perfect teamwork ended abruptly, and the coyotes fought for the meal just killed. When the fight was over, one of the coyotes left to do some supper hunting on his own. John T and I had been so engrossed in watching that neither had drawn carbines from our saddle scabbards to kill the coyotes, and we were glad that we didn't, for we had been treated to a show rarely witnessed by men, and the players deserved to live for putting it on for us to watch.

The camp was never without quail whenever a meal of them was wanted; the supply was inexhaustible, and we did not shoot or trap even one. Our hunters were as dependable as hot days on the desert in September. The camp had two cats, both females, and although we never saw a tomcat, nor heard of any within miles, our cats became mothers, each with a sizeable family. For some reason known only to cats, both selected John T's and my bedroom for motherhood and its exacting duties. No matter how many times they and their kittens were dispossessed, they always found some means of getting back. In the end we relented and cut a hole in one of the side boards where snakes could not come in and we hoped skunks wouldn't, put a canvass over the opening, and let the mothers come and go as they pleased. It was a happy solution to our cat-family problem, and the returns were manifold.

In the wash below our tent house, within a hundred yards, was a small spring to which quail came for water, and to roost at night in the Palo Cristi trees near it. The mothers were lazy cats and confined their hunting to as near camp as possible, and the quail in the Palo Cristi were nearby and an easy prey; so after dark they would go to the hunting grounds and return each with a quail for its kittens. John T and I were unwilling to permit cats to deplete the quail supply without our participation. When one of the mothers came through the opening with its catch, we appropriated the quail. We had been more than lenient with the mothers, and they owed us something for not depriving them of their broods. Had the cats been at Big Stick and seen the results of feline multiplication and our successful effort to thin it down, they would have known how generous we really were.

The mothers would cry around for a time, begging us to return the quail. When they understood the futility of it, they would go out and return shortly with another. We never took more than we needed for dinner for four, and when those requirements had been met, the mothers were allowed to keep the surplus for their offspring or themselves. When the quail harvest first got under way, the bird was a delicacy for about a week of dinners; after that, we wanted quail less often, and by the time a month had passed, none of us cared if we ever saw or ate another quail; in fact, we preferred never to have them mentioned. The mothers probably were happy to go on a routine of catching fewer quail, but they never tired of the diet and persisted in reducing the quail population even after their offspring were weaned.

When John T and I had enough of aimlessly riding the hills

all day and some of the night to avoid the obnoxious tenderfeet, whom we couldn't avoid meeting anyway, we gave up, called it deep enough, and abandoned our wandering expeditions. October was coming and our patience nearly exhausted after six months of waiting for Dilthey to come out or to make up his mind to do something with the Copper Belt. John T and I decided that if Dilthey did not show signs of life soon, we would call it deep enough, and I wrote him to that effect.

While we waited for Dilthey to reply, we resumed our regular evening visits to the goat ranches and gave more attention to our ranch-girl prospects whom we had sadly neglected. We spent a small part of our time with them looking at Halley's comet. Daylight hours passed at Constellation where John T and I dealt at Wren's twenty-one table and faro bank and at the same time got a good look at the nature of mining promoters, tenderfeet, and hangers-on at mining camps. All but a very few of the players at the tables were from the East and as losers were the worst sports I have ever met. Even Sweet Caporal stood much higher in my estimation.

Usually we got away from Wren's in time to reach the goat ranches for supper, but on a few days we were delayed when some new tenderfoot blood showed up and tried to show how easy it was to clean out the house at one of the tables. Occasionally there would be heavy arguments that often ended in brawls that were a result of opposing claims on the merits of the Cook and Peary battle as to which had discovered the North Pole. We made it hard for the tenderfeet to take Wren for a cleaning, but we could do nothing about Cook and Peary and let the fight that developed from the arguments take the course that nature intended.

When finally a letter came from Dilthey, he told me to come to New York for a conference. It took little time to get under way after first sending Ted to the Big Stick to get his hands blistered and his muscles sore learning mining the hard way under Sully's tutelage. Ted had less than a year to wait before entering Colorado School of Mines, and I wanted him to learn as much as he could about what hard-rock working stiffs had to do before he absorbed too much technical learning. I gathered from Dilthey's letter that nothing would be done at the Copper Belt, at least for some months, and John T was free to divide his time between the Wren and the goat ranches. The Wren was taking care of itself, and I knew that the prospects at the ranchhouses would receive most of his expert attention.

My journey was slowed down a few days by stops at the Big Stick, and at Prescott to see Tom and O'Sullivan, after which I drifted to Chloride. The mines were both going fine: the mill at the Big Stick was turning out profits for Tom, and that at Chloride had been in operation a couple of weeks and the first shipments of silver concentrates were on their way to Garfield to be smelted. C. W., who had come out when the mill got under way, was satisfied that Jack knew what he was doing, and he joined me on the final leg of my trip to New York.

SKELETONS IN A CABIN

C. W. and I parted at the Grand Central Station, and I went to the Copper Belt offices on Broad Street to meet Dilthey. The company offices were a plush affair, and Dilthey, director, president, and largest stockholder, was holding forth behind a smoothly-polished desk that had samples of gold ore on felt mats in front of the leather cornered blotting-paper holder on which he was leaning.

Dilthey was patently a promoter, and I already knew that he had controlled the company since the Wickenburg property was transferred from the Denver promoters. The company had five directors, Dilthey and four others, two of whom were his friends. There were also C. H. Peckworth, a successful building contractor, as the treasurer, and S. Bloom, a retail merchant, as the vice-president. Both were men of integrity and high principles.

It took a few minutes to repeat to Dilthey what I had written in my reports, which, he told me, he had not shown to the directors, nor had he made their contents known to them. Dilthey and I did not agree, and his lack of experience in mining was manifest when I told him the Copper Belt was a prospect and that the "mill" was an empty shell. He became emphatic when he told me that he had been to the property several times and that it was a mine and not a prospect. He insisted that all that was necessary to make the mine pay dividends was to take ore from the shaft and run it through the mill. It was all over but the shouting when Dilthey asked what I intended doing with the Wren claims and I told him it was none of his business, the property was not his nor did it belong to the company.

Dilthey did not let the matter drop there but produced an option for the claims signed by Wren and also a deed, which had been recorded. The option called for a deed in escrow to be delivered when payment for the Wren claims had been made. The deed "in escrow" had been recorded, and on the face of it, the Copper Belt was the owner of the claims. Something was not on the up-and-up, and where to place the blame did not matter. I had had enough of Dilthey and the Copper Belt anyway. It took less than a minute to write my resignation, hand it to Dilthey, and leave him to explain to the directors as best he could.

The rest of the day was spent with C. W. preparing a report for the group that had backed him in the Chloride Mine venture. C. W. had done a real job for them. He had developed a mine, built a milling plant to treat the ore, and earned enough to return the original jackpot, and in addition cached away a reserve to cover any emergency. When we took the train that night for Boston, where five of the group lived, C. W. looked forward to handing a check to each and letting it be known that more was to follow.

When the meeting in Boston was over, two of the group, Herman Hormel, one of the big guns in behind-the-scenes Republican politics of Massachusetts, and Doctor J. E. Meyers, an outstanding physician and surgeon, approached me with a problem on which they needed help. The problem was double barrelled—a lost family from which nothing had been heard since 1878, and a lost mine. The scene was "somewhere" in Nevada. Would I trace the family and find the mine? I was glad that the territory was narrowed to one state and didn't embrace the entire western mining area! Even at that Nevada was no backyard, and it looked as if there was a lot of territory to be covered. I told Hormel and Meyers that it would be easier to find the family than the mine.

"Lost" mines were mostly imaginative and invented by some prospector who was looking for a grubstake. His tall tale was usually so plausible that a legend of a lost mine came into being. None that I had ever heard of had been found. Hormel and Doc Meyers were anxious to have help, and, because they were friends of C. W., I agreed.

Hormel took us to his office and brought from his safe letters and papers that he thought would be of help in finding both the family and the mine. What we saw and what Hormel had to tell convinced C. W. and me that this was something different: the family had disappeared without trace, and there was a real and not a dreamed-up "lost" mine. Hormel told us that Joshua Ward, some sort of a relative, had gone west in the spring of 1876; with him were his wife, Abigail, and their five- and eight-year-old daughters, Sarah and Phoebe. Letters had been received irregularly until August 1878. The last one, written August 5th, bore an August 12th, 1878, postmark of Cherry Creek, Nevada.

Early inquiries as to the whereabouts of the Wards through post offices and law enforcement agencies in Nevada had failed to develop information other than that no one had seen or heard from them, and it was assumed they had left the country. Several of the letters from Abigail had information which, although vague, would be of help. In no letter, however, was there any mention of moving elsewhere; in fact, there was every reason to believe that things were going so well that a move would not be contemplated. One letter contained rough sketches of their log cabin and of its interior; in another was a crude map showing the location of the cabin with an arrow pointing south to Hamilton, another to Cherry Creek on the east, a third to Humboldt up north. The distances to the towns were not mentioned but below an arrow pointing to west of Hamilton was written, "Eureka, six days." The letter containing the map told of the journey to Eureka with the first load of silver ore, and in another to Cherry Creek for supplies. Both trips were with bull-teams, and Abigail wrote of their slowness in covering the distances they must travel, and that after two or three more trips to Eureka with ore they would have enough money to buy two teams of horses.

The letters gave us something more than wild guesses to go on. A six day bull-team journey to Eureka marked the distance as about eighty miles, give or take ten either way, and Cherry Creek nearer than Eureka because supplies would not have been purchased there were the distance greater. It was a relief to know that the area over which the search was to be made was not the one hundred and ten thousand square miles I had at first anticipated, but only four hundred.

A few small pieces of ore that had come with one of the letters would help to identify Joshua's hole in the ground. But that hole was going to be one hell of a spot to find, much harder than the cabin.

I doubted that ore came from anywhere near the cabin, because

Abigail, in one of her letters, told of Joshua's leaving for two weeks to mine enough for a bull-team load. Had the mine been close-by Joshua would have gone out in the morning after breakfast and returned in the evening. I guessed that it was distant about a two hours walk or more, which spotted it within the circumference of a circle with a ten-mile radius. At the least there were almost one hundred square miles to cover to find the mine, which I might do with abundant luck.

The descriptive sketches of the log cabin were reassuring. If the cabin remained standing after battling the desert storms for over thirty years, it could not be mistaken for any other. The cabin was described as being in a small and narrow basin-like valley at the upper end of which was a spring and a few cotton-wood trees. Joshua had built a road that came over a ridge into the valley from the upper end, went past the spring to the cabin and on to a field where the oxen were grazed. A hundred feet in front of the cabin was a shed under which the oxen sheltered and an outhouse in which deer and antelope meats were smoked. The cabin was L-shaped, the livingroom was the perpendicular and the kitchen the base. There was a door which gave access to the livingroom, but no windows, so that only one opening had to be watched when Indians that hunted in the area came around foraging. The cabin was shingled, and the roofs of the bull-shed and smokehouse mud-covered and strawed-over.

I had no illusions that the search would be a simple one. Winter would soon close in and weather might give trouble, but in addition the country was mountainous with elevations from five to ten thousand feet. I had been over similar country, and the mental picture I had of what had to be done to find an isolated cabin in an otherwise uninhabited area of hundreds of square miles was not a pleasing one. Hormel wanted me to get started as of the day before. Although winter was against my chances of finding the cabin and claims so late in the season, I wanted to get west again as soon as I could. If the job couldn't be done before winter set in, it could be done when the snow was gone in the spring.

Since I would need help, I sent a telegram to Ted to get to Ely as fast as he could. He was to buy a Ford, or anything else he could get his hands on, new or old so long as it would run, take a lot of abuse, and not break down too often.

I was in Ely eight days later, and Ted was waiting with a new

Ford, spare parts for it to carry along in case we broke down, extra tires, tubes, drive shaft, axles, and similar items. On the rear he had had a rack built to hold two cases of gasoline—twenty gallons—and enough oil and grease to put a slick around a battleship. Ted did not know where we were going or for what reason, but had filled the car with enough to carry us a month. Bedrolls and everything Ted had bought were loaded and waiting when I arrived, including our rifles, shotguns, and enough ammunition to fight the Spanish war over again.

We made Cherry Creek from Ely the night of my arrival. We spent the following day making inquiries and studying every map the curious citizenry would let us lay hands on. When evening came, we put together the pieces of information we had gathered, came up with what we thought was the best locality to make a start: the northern end of the valley and lower ranges of hills between the Butte and Diamond mountains. No road shown on any of the maps we looked at went within miles of where we intended to make the search. We hoped that luck would be with us and that one of the old unused roads would show up and give us a chance to drive closer than we thought we could.

Dawn was breaking when we left Cherry Creek. There were few traveled roads, and most of those mere marks on the landscape through valleys, up passes, and over the mountain ridges. There were no settlements or even isolated prospectors and their cabins. Game was plentiful and there was water, if one knew where to look—not in the low places but high up the hillsides or in small valleys near the tops of mountain ranges. On the desert, it was an axiom that one dug for firewood and climbed for water. Many a poor devil of a tenderfoot and prospector, too, who should have known better, died of thirst because he dug for water in dry stream beds or in the bottom of a valley.

By mid-morning we noticed the marks of a road, long unused, that crossed the north-south one we were on. It was indicated on the older maps as the hard but shortest way to Elko, passing between Ruby and Franklin lakes through the Harrison Pass and over the Ruby Mountains. However, it cut across the country farther north than we intended to go, and the only thing it was good for was to give us a landmark and let us know where we were. After we had sized up the situation, we backtracked twenty miles looking for signs of any kind of a mark that would mean that a road or trail had once been there. We were backtracking the fourth time and driving the Ford at a walking pace when Ted thought he spotted what we were looking for but had given up hope of finding.

We stopped, got out of the Ford, and followed the tracks. They were unquestionably an old and now almost obliterated road with sage and brush growing on it. It was headed in the right direction according to my calculations. We returned to the car and in it followed the tracks, often getting out when the marks disappeared, and back in again to start once more when they were again located. We were finally stopped where water from a cloud-burst had long ago made a deep cut. Rather than spend a day digging ramps in and out for the car to pass, we got out and started with thinned out bedrolls, canteens, food, and other necessities. The rest of the way we would make on foot.

Instead of becoming worse, the old road marks became clearer as we reached into the foothills and started up a long pass where it had been necessary for the maker of the road to cut away slopes and move rock so that a wagon could go through. The going was so rugged that we were slowed down, and as the sun was about to set, we were at the top of the first ridge twenty miles from the Ford. We stopped to take a breathing rest before going on and saw in the pass, a mile away, the cabin we had been looking for. We were not surprised that the ancient road led to it; we would have been surprised only had it not.

Darkness came fast, and when we reached the spring a hundred yards up the valley from the cabin, we camped there and let our curiosity wait until morning. We were out of our bedrolls with the first light of morning and looked around. Abigail had described the cabin and its surroundings so well that it seemed to me that I had been to it before. Abigail had missed oniy one detail, a low rim-rock ridge back of the cabin that sheltered it from cold winter winds.

Joshua had built well. The buildings were all standing with little evidence of the weather and storms that had pounded at them for thirty years. There were signs that deer, antelope, and a few stray cattle had watered from the spring and had grazed nearby, but nothing to show that any white man or Indian had been there since Joshua left.

Breakfast out of the way, we went to work in earnest to inspect the area. Clouds were gathering and we knew there was no time to waste. A storm would be on us soon, and before it struck, we wanted to be back over the ridge and in the Ford on our way to Cherry Creek. Our luck had been more than good, and to press it against the first winter storm would be tempting fate too far.

Halfway between the spring and the corner of the kitchen part of the cabin was an old ore wagon, its wagonbed resting on the huge wheels sunk in the ground with only a small segment of them with their wide iron rims above the ground. The wagon-body had been filled with ore, but the sides had long since broken and spilled the silver ore with which it had been loaded to the ground. The weight of the ore, and the water-soaked ground, had been the cause of the wagon's sinking. A bull-yoke with the U-collars gone lay near the wagon tongue and was partly buried. The weathered, grey-white bones of two oxen lay in the shed, skulls broken between the horns as if by the flat of an axe, each unmistakably killed with a single blow.

We looked at everything outside before going to the cabin. When we reached the doorway and saw arrowpoints with broken shafts in the log wall on either side of it, neither had an over-whelming desire to go in. The door would not open, and only the upper part could be broken sufficiently to make an opening for us to crawl through. Once inside we saw that the difficulties with the door resulted from its having three bars across it. There was little light coming through the opening in the door, but we could distinguish an overturned table in the center of the room and four chairs on the floor beside it. Against the wall to the right was an organ, and against the wall nearest it bookshelves; beside it was a home-made desk and against the wall opposite the door two beds. There was a deep accumulation of dust on the floor, hardened from water that had come through the leaking roof shingles, and similar hard-crusted dust covered everything.

I broke the hardened dust near the doorway, removed the bars, and opened what remained of the door to let more light in. Near the wall beside the doorway was a human body outlined beneath the hardened dust that covered it. On the bed near the fireplace was another body, and in the corner between it and the fireplace were two smaller forms. We broke the crusted dust from the bodies to find browned, dried flesh shrunken against the exposed head and hand bones. Disintegrating clothing had glued itself fast against other parts of the shrunken forms. The skulls of Abigail, Sarah, and Phoebe had each been crushed by a single blow with a wide dull weapon, similar to that which had destroyed the oxen. Joshua's body told nothing as to how he had met death, until we turned it over and found the broken shafts of four arrows protruding from the back, broken at the body no doubt after he had barred the door and fallen to die where we found him.

There was no way of knowing what had actually happened, but Ted and I reconstructed events as best we could from what we found in and out of the cabin. Indians must have sneaked into the cabin while Joshua was outside and killed Abigail and the children. Joshua had made for the cabin, but as he went through the doorway, he must have been hit by the arrows, which killed him; he had had time and strength to bar the door before the Indians got to it again. Why the Indians did not break in and loot it was a question that only Indians could answer, but it was obvious that no man had been in Joshua's cabin from the day he died until we arrived.

We moved the bodies of Joshua and the two girls, laid them on the bed beside Abigail, and covered them with the disintegrated remnants of bedding from the other. It was almost noon when we completed the unpleasant chore. During the time we had been busy in the cabin, snow had begun to fall, and we wanted to leave before the storm let go in earnest and gave more trouble than we cared to be mixed up in. Before pulling out we took everything from the desk and from under a rough piece of board on the bookshelves copies of the Cherry Creek newspaper, *The Independent,* bearing dates from May 18th to August 10th, 1878. After repairing the broken door, we headed into the storm. There were many things that remained to be done, but we would return to finish our job when winter was over.

The storm marooned us in Cherry Creek, which we made at dusk the day following our departure from Joshua's cabin. It took longer to return to Cherry Creek than it had to find the cabin. Less than half way on our return to the Ford the snow was so deep that we couldn't follow the old-road tracks. We couldn't see any of the landmarks we had spotted on our way in, and we were completely lost. We made camp under a low juniper, threw out bed-tarps over it for added shelter, and waited for the storm to give up. In the morning the clouds broke temporarily, and it stopped snowing long enough for us to see the Ford six or seven miles away in the valley below. Our troubles were not over after we reached the Ford, for it took everything we had to get it through the more than a foot of wet snow, on the road-trail that we frequently lost altogether and we had to retrace until we found it again.

The townfolk of Cherry Creek welcomed us and warmed our cold-stiffened bodies with the best red-eye in town. The next day at Cherry Creek I immediately sent a message off to Hormel and Doc Mayers to let them know that we had found the cabin and Joshua and his family dead, killed by Indians thirty years earlier; that a storm prevented our remaining to complete the work we had come to do, and that I would return in the spring as soon as the roads were passable; meanwhile, a letter was on the way as well as some personal effects we had taken from Joshua's cabin.

During our stay at Cherry Creek waiting for the storm to let up, J. W. Walker, a mine-owner and operator of properties near-by, took us into his home and with us visited those mines that we could easily reach. Among the mines we visited were the Imperial and Exchequer, both silver mines, and it was not many years later that I was consulting engineer for the operators of these, and for other operators of mines in the district.

The townfolk were friendly at the time Ted and I were on the lost-family, lost-mine undertaking, and also in 1914-1915 when I returned to do some engineering work at the Imperial and Exchequer for Jim McCoy who was operating the mines with his partner, Shurtliff. In 1919 Walker was operating, and I returned the following year to become again active at the Imperial and Exchequer. Within a few weeks of my rearrival in 1920 the friendship thing was a different story—with the women of the town! There were open threats to have me tarred and feathered and ridden out of town on a fence rail.

Cherry Creek had lost its sporting houses, and the girls of the line had left town and gone to Ely, which was more active. On paydays many more than half of the hard-rock stiffs, working in mines near Cherry Creek, would go to Ely to spend what they had drawn in pay. The road to Ely was nothing on which to stage a Vanderbilt Cup race, and a lot of the stiffs landed in the hospital because of auto wrecks, usually on the returning trip, and occasionally one would go six feet underground.

Three or four days was not an unusual time for the stiffs to be off the job, and some did not return for a week or ten days, others not at all. Many who did return were unfit for work and returned to Ely to be treated by a doctor. Keeping a full mine crew was a problem, but Cherry Creek had its problems, too, created by the younger and more attractive stiffs who did not go to Ely. Not a few who remained in town after paydays became husbands by the shotgun route.

I had no interest or concern whatever in the morals of Cherry Creek or what happened to erstwhile virgin daughters, but I was concerned in having a full crew of hard-rock stiffs and getting ore out. Cherry Creek did not agree that my solution of the problem was what it wanted, but it suited the crew and made money for the operators I was doing the engineering for. Cherry Creek benefited, too, for money that formerly went to Ely was spent in town, and the virgin population did better than remain static. It increased.

Wren's rest home for overworked girls gave me the idea, so I made arrangements with important personages of Ely who had close and intimate connections with dance halls, parlor houses, and cribs. Nevada was wide open and almost anything was permitted; however, before completing my arrangements I consulted Judge Dan MacDonald of Ely. I was associated with Dan in mining ventures after he discovered the Ruth mine near Ely, named it for his daughter, and sold it to the Nevada Consolidated Copper Company. I wanted Dan's opinion before undertaking my new venture.

When the necessary arrangements had been made for six girls to come to the Exchequer, I fixed up one of the bunkhouses with rooms for rent, set up a small bar, put in a piano, and had the floor polished for dancing. Soon the rooms were rented, the girls paid for room and board, and ate with the stiffs in the cookhouse. No stiff was permitted to drink within eight hours of his going on shift, and even so his drinking was limited. None of the girls was permitted to charge anything for the relaxing entertanment they might offer. There was no restriction, however, on the stiffs' making monetary gifts so that diamonds could be purchased should any of the girls want to waste her money on such baubles.

The plan worked and the mines gave up enough ore to satisfy me, which was enough to satisfy anyone. Furthermore, as the necessity for using shotguns decreased, the irate matrons of Cherry Creek lost interest in my activities. The experiment was a valuable one, and I repeated it as a part of my mining operations at mines where camps were distant or isolated from a large town.

At one, where snow closed the mine in for three to four months each winter and stiffs pulled down their time after the first snow, I equipped the cookhouse with waitresses to the number of each man on the payroll and provided quarters that would satisfactorily take care of the situation. There were no shortages in working stiff personnel and the plan paid off beyond expectations. The only reduction of the payroll came in the spring when some of the stiffs and waitresses were lost through the marriage route, and with no shotguns necessary.

When the storm let up in Cherry Creek, Ted and I headed for Wickenburg by way of Las Vegas, Chloride, and the Big Stick.

We pulled into Las Vegas, cleaned up, and drifted into town to put on the nose bag. After what Ted and I had found in the cabin, we were in no mood to make a round of the saloons where we would most certainly run into a bunch of stiffs we knew which would mean a celebration we did not care to have. We missed getting mixed up in a celebration, but we did run into Jack Coyle and Bob English, old Goldfield hard-rock stiffs who were herding a short, stocky character into one of the Chinese restaurants where, because it was run by Chinese, the food was good. The character was introduced as "Pete." No one on the desert ever used his full name, Peter Buol, because he was the kind of man no one could address as Mister or Peter Buol after he had talked with him five minutes. Pete was a man of many parts—politician, mining engineer, promoter, or anything else, so long as the dollar Pete could earn was unquestionably an honest one, clean and uncorrupted. He was one of the few men of high and undeviating principle that I ever met.

Jack and Bob were looking for a winter grubstake and planned to soften up their intended victim with the best food in town. Pete, I learned not so long afterwards, played the part of a sucker for any prospector who was looking for a grubstake, which was mostly before cold weather set in and a stake was needed so that the stiffs could buy supplies to hole up for the winter. It was not as easy to get the grubstake as some of the stiffs thought it would be. It took from three to four hours to talk Pete into giving it to them, and any who failed in the time test went away empty-handed. Pete figured if the prospector needed money badly enough or thought he knew where he could locate a good claim, he would take the time to convince him of it. More lacked patience than had it, and these did not get the grubstake they were after from Pete.

Pete had no illusions about grubstaking and didn't expect to get anything from one, but he secretly hoped that someday lightning would strike and make a few dollars for the prospector and for himself. Pete quit the grubstaking business when Lovell, a hard-rock stiff turned prospector, showed up with a new gimmick. Lovell wanted a lot more money than any prospector had struck him for. He didn't have to talk the regular three or four hours after he explained that he wasn't going to look for gold, silver, or any of the metallic mineral ores, but wanted to find borax or one of the non-metallics. Pete fell like a ton of brick and brought me into the thing together with several others of his friends. Lovell found something, all right, and Pete went to a lot of trouble to sell it. I was not around at the time Pete made the sale to the West End Chemical Company, of San Francisco, but I got a check from Pete for my share, $31,250, which was a lot more than the two-bits I had put in.

It took Jack and Bob the regulation three hours to get their grubstake, while Pete ate two full meals instead of the one the stiffs had expected him to consume. While the drilling and blasting was on, Ted, Pete, and I made a friendship that was to last more than thirty years, until Pete went up the canyon to join others of the desert breed. Pete was not born to the desert. He had come to Las Vegas as an employee of the San Pedro, Los Angeles and Salt Lake Railroad, when the road was being built and Las Vegas was a construction camp and a place to get water. The desert adopted Pete the first day he went on it, and Pete and the desert both were well rewarded.

Pete was the "Father of Las Vegas" and went about making a real place of it before anybody realized it could ever be more than a whistlestop or a watertank settlement. The railroad brought some water in for their crews from Caliente, and Pete knew that water must be developed, and was the first to discover there was an abundant underground supply fed from Mount Charleston and the Spring Mountains. When the town began to grow out of its boots, Pete was mayor, and I did my two-bits worth helping him to get water to make his dreams come true. Pete continued to work to put Las Vegas on the map when he became a state senator, with the result that it grew beyond the water supply, and Pete had me help again to get that difficulty solved.

Water was not Pete's only problem when he became Las Vegas' first mayor and decided to make it a real city. The San Pedro, Los Angeles, and Salt Lake Railroad, which the Union Pacific later took over, selected Las Vegas as a division point. As a result, a large number of working stiffs were required to build the yards and to construct a station and loading platforms. When the working stiffs came in, a lot of undesirable characters followed and set up establishments that would take their pay from them. Most of the girls who came had been run out of mining camps or boom towns, as had the men who pimped them. There were gamblers by the dozen, not interested in honest games or satisfied with regular house percentages, who used loaded dice, rigged roulette wheels, dealt from the bottom of the deck, or used "cold" ones.

Pete knew that if disorder was allowed to continue and take root, no prospector, working stiff, or cattleman would come near it, and if they didn't his dream of a great city would go over the waste dump. Pete cleaned out the undesirable girls and their pimps, and established the line in a block near the center of town where "decent" girls could live, and where their activities could be easily supervised. His next move was to clean up crooked gambling. This took a little longer than it had to get the line straightened out, but he got what he wanted and Las Vegas was on its way. There were always enough girls on the line, and gambling; but the line was clean and the gambling honest.

Pete's troubles were not over, however, when he had cleaned up Las Vegas. Many of the undesirables had gone to Searchlight, some to Jean and Goodsprings; not all of them had left the area, and Pete wanted not only Las Vegas clean but Clark County as well. It was a tough job, but Pete, who by that time had become a state senator, kept at it.

Probably the final clean-up would have taken longer had not Paul Koski, who was a skinner for me at the Boss Mine, been killed in a stud game in Goodsprings. I was sitting beside Paul when it happened. Paul was a big winner, and the house dealer apparently made up his mind to call it deep enough and put an end to his winning. Paul began to lose. On his last hand, Paul held four kings, but during the deal he had spotted the dealer pulling cards from the bottom of the deck. He said nothing until the hole cards were turned. The dealer had four aces, but Paul reached for the money on the table, told the dealer he had cheated, and started raking in the cash. Without a word the dealer pulled a gun and killed Paul.

The local deputy sheriff refused to arrest the dealer, and the justice of the peace would not make out a warrant of arrest that I wanted to swear to. The following day I went to Las Vegas, saw Pete, and told him what had happened. A warrant for the dealer's arrest was issued, he was arrested, tried and acquitted when witnesses testified that he had killed Paul in self defense. Pete went into action after the verdict was rendered, and so did the sheriff of Clark County. Within a few weeks Clark County was rid of its undesirables, and Pete went on to see his project, and dream, made a reality. In addition to the sheriff, Pete had Fred Hess, another of Las Vegas' mayors, on his side helping him. Las Vegas became the metropolitan center that he dreamed it would be.

Pete and I drifted in and out of a lot of the desert mining camps in the years that followed our first meeting. Sometimes Ted was along and John T, too, but not as often. John T was usually blasting away at some mine to make it give up what was in it so that the owners, for whom Crampton and Crampton were the engineers, would get something for the dollars they put up, and Crampton and Crampton the credit.

Above: *Joshua Ward's cabin near Ruby Lake, Nevada, May, 1910. Photo taken by Doc Mayers about six months after the bodies of Joshua and his family were found (they had been killed by Indians in August, 1878) and after a fence had been built.* Below: *The Cherry Creek, Nevada,* Independent *of May 18 to August 10, 1878, found in Joshua's cabin in November, 1909.* Crampton Collection.

THE INDEPENDENT.

VOL. 1. CHERRY CREEK, NEV., SATURDAY, MAY 18, 1878. NO. 38.

INDEPENDENT

A. V. HOYT,
Editor and Publisher.

THE INDEPENDENT.

CHERRY CREEK, NEV., SATURDAY, AUGUST 10, 1878. NO. 50.

A. V. HOYT,
Editor and Publisher.

THE SONG OF HARD TIMES.

J. F. WILLIAMS & CO.
Lumber Dealers.

Mining Timbers

W. H. FARRAR,
ASSAYER, MAIN STREET.

Wells, Nevada, June, 1910. Doc (Dr. J. E.) Mayers and Herman Hormel en route to Golden with Ted (Crampton) and the author. Crampton Collection.

Mines rooters before Mines-Colorado College football game. Mines won! Crampton Collection.

Uranium Mine Camp 2, Dolores River, Colorado, fall, 1913. The author is riding the leader of the mule pack train with supplies for the cook house, which is on the right; behind is the cave-cliff dwelling. Crampton Collection.

Uranium Mine, Dolores River, Colorado, fall, 1913. Cook driving on return to the mine, cook's wife seated beside him, author sitting on the load. Crampton Collection.

The rapid transit of the early 1900s, Crampton driving. Crampton Collection.

Uranium mine, Dolores River, Colorado, in the fall of 1912. Pete Lund (left) and Ed Moore are camping for lunch between the uranium mine and Dolores. Crampton Collection.

The main road to Cherry Creek, Nevada, April, 1910. The main part of the town was on the left and did not show in the picture. Crampton Collection.

The author on his wild-herd pony, Nig, at the uranium mine on the Dolores River in the fall of 1912. Crampton Collection.

The Vanadium Saloon at Newmire (later renamed Vanadium), Colorado, in the spring of 1913. The ore reduction plant of the Primos Chemical Company is in the background. Crampton Collection.

Primos Chemical Company ore reduction plant, Newmire, Colorado, spring, 1913. Narrow-gauge railroad train in the foreground. Crampton Collection.

Primos Chemical Company ore reduction plant, Newmire, Colorado, summer, 1912. The author, holding a sign reading "SMILE! 70 TONS," and H. E. (Slim) Willis. Crampton Collection.

GRIZZLIES CAN FLY

Ted and I remained a few days in Las Vegas and were on our way again. We made a stop at the Chloride Mine, saw Tom and O'Sullivan in Prescott, then drifted to the Big Stick before heading for Wickenburg. The Chloride Mine had settled down to a production that could be depended upon for several years, and Jack had settled down with it and was soon to be married.

The Big Stick was doing well, but it would not last much longer, and Tom was unhappy about it. As was typical of many desert mines, the ore had not gone deep; the last hundred feet of the shaft was in barren ground, and the faces of drifts in both directions from the shaft had gone beyond the pay ore-shoot. A year later when I called her deep enough and shut her down, Tom yelled his head off, and I had to take a pick handle to make him see the light. I wasn't going to let him crowd his luck and lose his shirt trying to find ore that I was certain wasn't there. Sully was more worried than was Tom, for he had given up the three-month-stiff idea and was worried about getting another long-time job when the Big Stick shut down. Sully had nothing to be concerned about, for Crampton and Crampton had enough work to be done to keep him occupied.

O'Sullivan agreed to do something about the Wren claims and advised me to keep hands off the dispute that was developing between Wren and Dilthey and to let him straighten the thing out. Dilthey never gave O'Sullivan a chance, nor did Wren. They battled between themselves, came to no settlement, and John T, Ah Yat, and I were caught holding an empty sack. Before we pulled out from the Wren, however, John T, Ted, and I took out every pound of ore we could mine, gutted the hole, caved the mouth of the tunnel, and called it deep enough. I did not see the Wren claims again until I went back twenty-five years later to find the tunnel mouth still caved as it had been the day we shut her down.

When we were certain that winter was over and roads open, we headed for Cherry Creek and Joshua's cabin. I wrote Hormel and Doc that we were on our way, asked them to join us, and told them that we would not enter the cabin again until they arrived. By the time Hormel and Doc reached Cherry Creek, a crew under the direction of Ed Devenney, who came up from Wickenburg to boss the job, had worked the road over to make it passable and built a fence around the level part of the little valley from above the spring to below Joshua's cabin. Ed had also supervised the purchase of supplies and had hauled them for the road gang and to the cabin.

With Hormel and Doc, we entered Joshua's cabin, placed the bodies in caskets brought from Ely, and shipped them to Boston for burial. After the bodies had been removed and the cabin cleaned up and made livable again, we started looking for Joshua's silver mine.

That Joshua had found a rich silver vein was beyond question. The ore that he had loaded on the now sunken-to-the-ground wagonbed was worth over a thousand dollars a ton, and Joshua had certainly mined and sold several wagonloads of it. In cleaning out the fireplace we loosened a flat slab of rock; behind it was a cavity in which we found over five thousand dollars in gold pieces.

When it came time for Ted to leave for Golden, Colorado, to get started in Mines, Hormel and Doc went with us as far as Denver, and then went on to Boston. In the years that followed Hormel and Doc came out to Joshua's whenever I was free to meet them there and they could get away at the same time, which was not often. Several times Hormel and Doc accompanied me on trips I made to mines on the Nevada and California deserts. They saw enough good operating mines to make them more determined to find Joshua's silver mine, and they were certain that someday they would stumble onto it. However, they never did, nor did anyone else to my knowledge.

When John T, Ted, and I reached Golden, I was as surprised as Ted to find the family there and settled in a large house. I was greeted with some warmth, but there still remained a barrier on which I could not lay my finger. I learned what it was soon enough: I was a working stiff, and the family lived in constant fear that I would not always do the things that were "proper" in the "best" families. Past experience apparently had cautioned them to beware. I am certain that the only reason they appeared in Golden was to see that I did not influence Ted to become a follower in my footsteps.

I was doing some engineering for Fred Himrod at the Lamertine mine near Idaho Springs, and John T had gone up to get things under way. It was a strange coincidence that I was doing the work that Lafe Hanchett had formerly done for Himrod, before Hanchett went to Utah for the Boston Con. The family had arranged with Himrod for Ted and me to get the mine job for Crampton and Crampton, and also some work on remodeling the power plant near Georgetown in which Himrod and Hanchett were jointly interested.

I made frequent trips to Idaho Springs, and on one of them John T and I decided to have a celebration. The celebration was well under way when I heard the Denver-bound train whistle and decided to return to Golden, disregarding the fact that I had not cleaned up and was dressed in digging clothes splattered with mine mud, candle grease, and everything else that could be picked up underground.

I was asleep when the train hit Golden and failed to get off, and the conductor wakened me as it rolled into Denver. I was a sad-looking mess and had not a dime on me. The celebration had taken care of whatever loose money I had, or possibly someone had relieved me of it; whichever it was, I was broke. The solution to the money problem was to reverse the charges for a phone call to the family in Golden and ask for temporary financial assistance.

When the relief party entered the Union Station and came through the aisle where I sat, it gave me one look and passed by as if a wild bull were coming up from behind. I watched while a second relief trip was discussed, and finally the route down the aisle was retraced, no doubt to make certain that what they saw was me. There was again no sign of recognition. I waited, amused, to see what would follow. It was a redcap bearing a piece of paper folded and weighted from within by three twenty-dollar gold pieces. The note read "Get some decent clothes," and contained an invitation to luncheon at the Brown Palace if I did. When the time came, I was properly attired, had gotten a check cashed for enough to repay the temporary loan of sixty dollars, and enough more to pay for the luncheon.

What might have developed into an "incident" passed without comment. Tensions eased, and by the time Ted was graduated, the four years spent in watching others of the engineering breed altered somewhat the family's conception of proper behavior.

I think that one of the reasons the family changed its opinion of working stiffs and mining engineers was because Mines students and graduates were real he-men. Mines students were never accused of being softies or full of prunes, nor were their school yells. I remember a football game in Colorado Springs against Colorado College before which the powers-that-be of that school requested that Mines students refrain from giving yells in which there was any resemblance to profanity. That stopped every Mines yell cold! There was not one Mines yell that could be given without violating the edict!

The Mines football team and the rooters who went with it on the train to Colorado Springs were submerged in sadness, a sadness greater than when the legendary Casey struck out. Silence during the game was impossible. It would assure Colorado College of victory if Mines rooters could not encourage the team.

Something must be done, and it was! Several brilliant minds concocted a new yell. The authors of the masterpiece, as I remember, were Doc Pierce and "Tuffy" Wolf, both players on the team, and Gordon Bettles. Gordon was in the Philippines when war came, the Japanese took him prisoner, and when he became ill and couldn't work, killed him. We detrained and went down Pike's Peak Avenue on our way to the Antlers. The procession was silent for a strained minute or two until some raucous person among the sidewalk spectators shouted some disparaging remarks about the team and its being beaten before the game started. Then it came:

"Rickety Rickety russ, we're not allowed to cuss; but damn it to hell, we feel so well, we really, really must!" It was the nearest thing to a yell full of prunes that Mines ever had. And then the rest of the Mines yells followed, with complete abandon. Mines won 17 to 7.

During parts of the years 1911, 1912, and 1913, when I was not in the field or making a hurried trip to Boston or New York on Crampton and Crampton business, I worked for the Colorado Geological Survey under Professor G. M. Butler, "Monte" to Mines students, in research on the clays of eastern Colorado. Monte, the author of an outstanding work on mineralogy, went to Corvallis from Mines, and later to the University of Arizona.

Ted and I designed and built several high-temperature gas-burning furnaces in which we tested and melted the clays. We had one clay, however, that refused to be reasonable and become viscous, melt to the unitiated, "two-forty-one" we called it. Heat never did much to it. Several times we thought we had it but in the end it licked us, even when we designed and built a furnace just to make it melt and behave like a clay should.

The clay testing and furnace work with the Colorado Geological Survey was to be the reason for my transfer from the Golden Company of the Colorado National Guard when it became the 115th Engineers of the 40th Division and in training at Camp Kearney. The transfer was to Chemical Warfare Service in Cleveland, where I was to construct a furnace for the manufacture of activated carbon for use in Army gas masks.

During a part of the time I remained in Golden, I studied the chemistry and metallurgy of rare metals under the tutelage of Dr. Herman Fleck, an authority on the subject. As a result, I went on a construction job, a mill to treat uranium ore for the recovery of radium, which the American Rare Metals Company was mining on the Dolores River in southwestern Colorado. Dr. Herman Fleck, "Monte" Bulter, and William G., "King Bill" Haldane, then acting president of Mines, were interested in the radium recovery process around which the mill was designed.

The mine manager, C. D. Heaton, at one time assistant in geology at Mines, was a good mining engineer and geologist but not as good at mine management. Heaton had not thrown his books away nor had he learned how to swing a pick or handle a muck stick, and he didn't know what it meant to do a day's work. Heaton was continually looking down the necks of his working stiffs and trying to drive them beyond the capacity of any of them to work well.

The last straw with the mill and hard-rock stiffs was the day Heaton sat on the rimrock above the mill excavation and counted every stick of muck each stiff tossed into an Irish buggy in one minute, and the number of buggy trips the buggy made each hour. From there out not one of them put in a shift's work or anything near what would have been done for a manager who knew what it was all about.

Heaton got himself into a mess at the Number Two mine camp two miles from the mill. Tents and canvas shelters were put up for twelve of the hard-rock stiffs, and the eight who were not taken care of in those makeshift quarters threw their bedrolls down and camped in one of the several cliff dwellings that had been built hundreds of years before by Indians beneath the shelter of low but deep sandstone caves. The cookhouse had as its rear end one of the cliff dwellings in which a German cook and his wife made their quarters. The cooks had a hell of a time getting meals for twenty mining stiffs on a two-by-three foot, three-hole, wood-burning stove that was barely large enough to cook food for four.

The entire camp was sore at Heaton, the cooks about their quarters and the stove, and working stiffs about their quarters and cold food. In addition to the living inconveniences, mice, rats, and skunks roamed the cookhouse and sleeping quarters with the utmost abandon.

Shattucky Bill Jordan, the mine foreman, who had been a labor leader in the Cripple Creek, Idaho Springs, and Telluride strikes, blasted an agreement from Heaton to build decent quarters and to get the cookhouse situation straightened up. I had met Bill in Idaho Springs, and he knew that I was a member of the IWW and Western Federation of Miners. He trusted me, and we discussed the situation at considerable length. I didn't want to see the men strike because Dr. Fleck, "King Bill," and "Monte" would have been hurt by one and perhaps their positions at Mines jeopardized.

Shattucky was willing to go along for a while. There was no dissatisfaction at the Number One camp, but had the Number Two camp stiffs struck the mill, Number One camp stiffs would have gone out with them. It was not until the cook and his wife quit that Heaton agreed to get a large stove from Durango and have lumber hauled in for the construction of camp shacks.

With the cookstove and camp situation cleared up, including the return of the German cook and his wife, I wired Pete Lund in New York to come out to work on the books and payroll records. The family had written asking me to get him a job at whatever work I could find. Pete was a ticket seller for the Frohman theaters in New York, and I had visions of another Sweet Caporal, but I took a chance.

It was late one afternoon, when I was going into Dolores with Ed Moore who had a livery stable there and who did a lot of teaming to the mine, that I first met Pete, a mile out of town. Pete had walked out from town on the road every one of the five days he had been in Dolores, to meet me the sooner, and to kill time. Ed Moore and I sized Pete up at supper that evening and decided, with reservations, that he would make the grade.

We pulled out of Dolores two days later and gave Pete something to do that would prove to us whether we were right in our estimation of him or not. Ed Moore was taking two saddle-broken pack mules back to camp with us, and we saddled one and told Pete he would ride it to the mine. When we made camp for the night near Bill Snyder's, regardless of how sore Pete was from the bumping he had gotten he made no complaint and helped water, feed and bed down the teams; after that he got firewood and helped prepare supper.

The next morning Pete went for his saddle, but before he could throw it on the mule to cinch it on, Ed Moore grabbed it from him, tossed it on one of the wagons, and told Pete to get up there with it and rest his fanny the remaining miles to camp. Pete was at least honest. As he climbed onto the wagon, he told us he didn't think he could have ridden the rest of the way, but he had been willing to make the try.

Pete was all tenderfoot to the stiffs at camp, and they gave him the works. Pete took it all and appeared to be having fun riding along with what he knew to be a first-class hazing. The stiffs soon realized that Pete had something in him and were ready to call it quits when some bright mind came up with the suggestion of a snipe hunt. Pete, who had never heard of a snipe hunt before, knew something was wrong with the idea but went along as if he suspected nothing.

One clear night when the moon was high and there was enough light to throw shadows, the stiffs headed for a box canyon a mile from camp where Pete was told a lot of snipe could be caught. Pete was stationed at the closed end of the canyon with a gunny sack in which he was to put the snipe he would catch when they were driven up the canyon by the stiffs who would frighten them into going to where Pete was waiting. Lighted candles were left on each side of Pete to attract the snipe, and to make them come close to him so that he could catch them with his hands. Pete had borrowed my twenty-five thirty-five to take with him and had a full magazine and a shell in the chamber.

When everything was set, the stiffs left Pete and went down the canyon yelling and beating the bushes to scare the snipe. Not long after the noises from the canyon quieted down, Pete saw two close-together lights with a shadowy form behind them coming toward him from a distance of about two hundred feet. Pete was scared stiff and took the twenty-five thirty-five and let go, then cleared the breach, got another shell into the chamber, fired again, dropped the carbine, and ran for camp.

I had not turned in when Pete reached camp, and because my oil lamp was still burning, he headed for my tent, barged in, and sat down, breathless. He had run most of the mile to camp, and it took him some time to recover enough to tell me what had happened. By the time he had gotten the story out I heard the stiffs who had taken him on the hunt returning, so I told him to take the spare bedroll and bunk in the tent with me. The stiffs had meant no harm, but the riding they had given Pete over the last few weeks had to be stopped. I thought I knew how to do it, so I told him not to show up for breakfast until everyone had gone on shift and to let me take care of things.

Before any of the stiffs left the cookhouse at breakfast, I mentioned that I had heard two shots and inquired who had been out shooting deer by moonlight. There was unanimous denial. Then turning around to look at the other tables, I remarked that Pete was missing and wondered where he was. No one knew. One of the stiffs who had been in on the hunt suggested that Pete might have decided to remain with his gunny sack until it was light enough to find the trail to camp. I told him that didn't make sense and ordered the stiffs who had taken Pete on the snipe hunt to go out, find him, and bring him back, on their own time.

When the stiffs returned just before lunch, Pete of course was not with them; but fastened to and hanging from a long sapling pole, carried by four stiffs, two at each end, was a huge mountain lion. The dead animal was as large as the biggest male African lion I had ever seen in a zoo, and but for the missing mane it could easily be mistaken for one. Pete had hit it between the eyes, probably with his first shot, the second in the spine; whichever it was, either would have finished it. It was a good thing for Pete that his aim was good and that he had killed the brute instantly. Had he only wounded the animal it would have torn him to ribbons.

Bill Snyder, a cattle rancher who ran several thousand head, maintained a summer ranch on the river below camp. Bill had sold the uranium mines to the group Heaton represented, and I wanted to get a map from him that showed where the claims were located. There were two routes from the camp to Bill's. The shortest was a two-mile-long trail that followed rimrock until it reached a narrow gully that cut through the sandstone from the rim to the river bench one hundred and fifty feet below. It could be covered easily by foot and on a sure-footed pony. The longer route was a four-mile road which crossed the river by a ford over which Bill hauled ranch supplies.

After lunch one noon I saddled Nig, a tough pony that had been cut from a wild herd and broken, and headed for Bill's. Nig could go anywhere, and I think that if I had tried him he would have climbed a telephone pole. When we reached the gully where the trail went down to the river, Nig whirled and pranced and refused to go, although I hadn't had trouble getting him to go down the several times when I had ridden that way before. It was a short half mile to Bill's, and rather than fight Nig, I decided to make it on foot rather than ride to the road and take the longer route.

I left Nig on the rim and went down, forgetting to take my six-gun from the saddle holster and transfer it to the one I had on the cartridge belt I wore. As I walked along the trail to the river through aspen and willows, I heard a sound not twenty feet away, stopped to see what it was, and discovered a grizzly bear cub having a fine time in some berry bushes. The cub turned, gave me one unconcerned look, and went back to the business of cleaning up his lunch of berries.

A grizzly cub meant that the old she was somewhere nearby; so I went back up the trail a hundred feet to think things over and decide whether I would backtrack and ride the long trail on Nig or make a wide sweep around the place I had seen the cub and so on to Bill's. I listened for sounds of the old she, but heard nothing except the cub at work on the berry bushes. Still undecided, I let out some rebel yells until the canyon echoed with weird and unusual noises. There was still no sound of the old she that I was certain would have come back to the cub had she been within hearing distance.

Looking around netted nothing more than had my rebel yells; there was not a sign of the old she anywhere. I decided that the cub had become separated from its mother and was strictly on its own. Bill had often told me that he would like to get a grizzly cub and see how well he could tame it. My chance to give Bill the opportunity he wanted was at hand. I could kill two birds with one stone, deliver a cub to Bill and get the map I wanted.

I removed my cartridge belt, took the cartridges from it, put them in a pocket, and eased over toward the cub, which was more interested in cleaning up berry bushes than he was in my approach. When I was close enough, I threw the belt around its neck, pulled the leather tight through the buckle, and had a wildly twisting cub on my hands. Not only did the cub twist and fight to get free, but it let out squeals that became louder with every step I took.

As I reached the river bank, another noise from down stream interrupted my journey. One look was enough—the old she had heard its offspring's squeals and was on her way, and so was I the moment I saw her. I was in a hurry to get away, but I wasn't alarmed, for I was certain the old she would stop and attend to its cub, find it was all right, and forget about the cause of its troubles. I was wrong about her interest in the cub. She passed it by without stopping to give it a sniff and held her course toward me, on a lumbering run.

I reached the base of the rimrock and started up the boulder-filled trail with the old she bear not more than a hundred feet behind. I thought that once I started climbing she would give up and go back to her cub, but there was no such luck. When I reached a narrow shelf thirty feet below the rim, the old she was within twenty feet and gaining fast. Climbing is a misstatement—she was flying as if she were an arrow shot from a bow.

I knew I was through. Thirty feet more to climb was out of the question. The old she would have me less than half way to the rim. Moreover, if I did have the doubtful fortune to reach the top, Nig, when he saw or smelled the bear, would be off for parts unknown and far away, dragging bridle reins against rigid training habits.

I made the only decision possible, to go along the narrow bench until I came to another and smaller gully cut through the sandstone rimrock and down this. I had gained twenty feet on the bench, and the old she was forty feet behind. Somehow I reached bottom without falling and took a look. The old she had barely gotten started, and I had a good hundred-foot lead. I put on all the steam I had, and when I reached the gully in which was the trail, I took a look to the rear and saw the old she just appearing and turning to run in my direction. She was three hundred feet behind but would close up a lot of that distance before I could reach the rim, if my heart and wind held out that long.

I made the rim and went into the saddle before the old she got up. Nig went into a dance that almost kept me from pulling the six-gun and letting go, without bringing her down. I let Nig have his head until I could draw my carbine from its saddle holster, and with Nig making a hell of a fuss, I checked the magazine for a full load, dismounted, and waited for the grizzly to come close enough so that I wouldn't miss and could let her have everything the magazine held.

I waited until she was within two hundred feet and gave her the full magazine, but she didn't drop and kept on coming while I ran as fast as I could, trying unsuccessfully to get more shells into the magazine. It wasn't necessary, though; she fell within fifty feet of where I had stood while I was using her as a target.

It took almost an hour to round up Nig, and I had a hell of a time getting him back to rimrock and to a place where I could go down to the river bottom again, a half mile from the trail. When I dismounted, I tied Nig to a sapling, stuck the six-gun under my pants belt, untied my rope, put it over my left shoulder, and carried the carbine. I wasn't taking any more chances. One was enough, and I wouldn't forget it.

When I reached the river where I had dropped the cub, there was not one cub but two, both busy at berry bushes nearby and unconcerned that I was around. When I had roped and hog-tied them, I went to the rim again and back to camp for help in getting the cubs out and skinning the bear.

There was a sequel to the bear cub capture. A cowhand in Dolores who had an unbeaten English pit bulldog boasted that there was nothing within twenty pounds its weight or near its size that could lick it, for money, marbles, or chalk. I wasn't interested in marbles or chalk, took him up on the money, and entered the smallest of the cubs against his English pit.

The cubs had become tame and friendly with the attention the stiffs in camp had given them, and although they had become lazy and fat, I was sure that in a fight instinct would become uppermost; if my guess was wrong, I would lose a lot of money. The fight was to take place in a twenty-foot square ring with sides three feet high. The pit bull was to be chained on a four foot leash in one corner, the bear cub in the one opposite. When the signal was given each was to be released and the fight would be on.

There was a crowd five deep around the ring by the time the fight was to get under way. The pit bull was wasting its energy tugging at the leash while it growled and snarled and did its best to break free to be after the cub. The cub, giving no attention to the dog, was begging food from the stiffs outside the ring, with its paws on the top board unconcerned with anything that couldn't be eaten.

The signal to "go" was given, and the animals were released. The dog gave a lunge for the cub, but the cub didn't move its paws from the top board of the ring. It was apparently interested in nothing but food. When the dog was within a foot of it, the cub turned, gave the dog a swipe with one paw and then the other, and the fight was over. The dog's hide was torn from the flesh on both sides and some ribs broken, and while it was being taken from the ring, the cub resumed its begging as if nothing had happened. The dog recovered in two months, but I was told its cowhand owner did not enter it in another fight.

A few days after the fight, Ed Moore and I returned to the mine with four wagons with trailers, all loaded to the limit that axles and wheels would take, with supplies for the winter, and the cub along as a passenger. It began to snow before we reached Bill Snyder's winter ranch to camp for the night, and we were off to an early start in the morning so that we would reach the grade from rimrock to the Dolores River before the snow was so deep that going down would be too dangerous. The road into the river canyon was winding and steep for the three miles of it, sandstone cliffs going straight up on one side and straight down on the other.

The drive was dangerous enough even in good weather, and before starting down each driver checked horses, harness, and everything on or about the wagons. The rim was reached in mid-afternoon, and Ed Moore was the first to make a start, followed by the two other skinners, with me bringing up the rear. Everything about my outfit looked all right until I checked the brake closely. The brake lever had held tight in the ratchet before I waved the last driver ahead to go on, but when I went up to the driver's seat and unloosened it, one end of the ratchet bar dropped off. When I examined it to see what was wrong, I found that the bolt holding it to the wagon bed was gone.

I looked around in the snow but did not find the bolt, then in the wagon for some baling wire. There was none. I searched for anything I could improvise to hold the ratchet bar in place. When finally it was obvious the search would net nothing, I cut off the trailer, improvised a brake from a sapling, and started on the trip into the canyon.

Rope that I used on the improvised brake broke under the strain, and the weight of the wagon made the two mule teams come back into their collars. The wheelers were all but caught under the wagon between the wheels. There was nothing I could do but hold on to the iron rails of the skinner's seat, give the mules their head, and let them do the work. A moment after the rope broke and the wagon went out of control, the mules were into their collars again and making it on a dead gallop. I knew the mules could make it if one of them did not fall down or a wheel suddenly leave its axle on one of the sharp turns. I was barely able to hold on, and the wagon sometimes skidded almost to the edge of the road into the canyon below. The only worry I had was getting to camp alive.

The mules made it all right, and the wagon, too, but the stiffs at camp had to lift me from the wagon seat, watersoaked from the sweat of fear and motionless because of nerves that had me frozen. Ed Moore went up for the trailer the following day while I remained in camp.

THE LUDLOW MASSACRE

My work was finished, the mill was complete, and I was debating whether to remain a week or two longer to watch it in operation or to return to Golden. The decision was made for me. One of the mill stiffs broke a leg, a compound fracture that would require a doctor to get the bone back through torn flesh and to set it before his removal to Dolores for treatment.

Pete was the first to offer to ride to Dolores for a doctor. I told him to saddle Nig, and to bring Doc LeFerge back with him when he returned. On the ride in Pete stopped at Bill Snyder's, the Wolfs', and other ranches, changed ponies, and told the ranchers to have fresh mounts waiting when he returned with Doc. Pete made a fast ride in, but coming out, Doc, who wasn't used to a saddle, slowed the return. He was about all in when he reached camp and rested an hour before he went after the broken leg, but Pete seemed as fresh as when he started.

Heaton had no sympathy for the stiff with the broken leg and raised hell with him for getting himself hurt and causing so much inconvenience and trouble. When a wagon and trailer went out the following day to take Doc and the injured stiff to Dolores, Pete and a dozen others went along. It was deep enough for them, and for me, too. I rode Nig out.

Dolores was overstocked with cowhands in for the Christmas holidays. Had it not been that I had a room always rented and waiting at the Hogan, Pete and I would have slept in the hayloft of Ed Moore's stable and livery. There was barely standing room at the bars of any of the saloons in town, but we managed all right at the Red Dog where the cowhands were only three and four deep in front of the bar.

In the Red Dog one could get a chance at the twenty-one games by patiently waiting until one of the players got too drunk to see the spots on his cards, and the lookouts politely, or forcibly, if necessary, removed them and told them to get the hell out and sleep it off. There was not much to do in town but to raise plain and fancy hell. The few girls who held out in the houses that would have been the line had they all been in one locality, had been supplemented by volunteers, but there were not enough to go around. When the ardor of the cowhands cooled off by waiting for their turn, they left and put their time in at other places, mostly the saloons.

There was not a cowhand missing from any ranch within more than fifty miles; they were so thick that the hitching poles across the main drag from the bank, Dolores Mercantile Company, and J. J. Harris, not to mention the five or six saloons, had enough ponies tied to them to horse every man of the fabulous Light Brigade.

The cowhands, when they were not visiting the line girls, were getting plastered and taking turns losing their money at the gambling tables, or spreading it over the bars. Whether they went broke or not did not matter, for their credit was good, although it would take most of them the six months to the Fourth of July to earn enough to pay off the debt and start the chain over again. The sheriff and his deputies kept good order, and whenever a cowhand got overloaded and thought he was a six-gun badman, he was taken in hand and thrown into the clink to sleep it off until the urge to show off wore off. The clink was small and the would-be Billy-the-Kids were so many they slept on top of one another. One of the town characters who spent most of the time in the clink for some petty breach of the law, or who had regularly to be taken there to sleep it off, gave the crowd in the Red Dog a lot of entertaining fun, for a time. Drinks were handed him without restraint, and, as a result, he became poetic and recited one after another of the familiar ribald ditties we knew, and a lot that we had not heard. After a couple of hours of recitations, the stiff stumbled out into the night.

In an hour he was back, as full of whiskey and poetry as ever, and a pocket full of money. Where he had bummed the money, no one knew, but he had enough to buy drinks for every cowhand and stiff in the house; for a fleeting moment he had attained one of the great ambitions of his life. When the drinks had been tossed down, he went to the center of the barroom, pulled colored chalk crayons from a pocket, drew a picture on the floor, and went into the routine of "The Face on the Bar Room Floor."

When the face on the bar room floor had been taken care of, a dozen hands standing at the bar each offered him a drink; he accepted them all, drinking one after another as fast as his adams apple would let them down. Into his last free drink he dropped some white powder; I thought that it was probably morphine or another of the easily obtained dopes that he was always taking, or taking the cure for. Whichever it might be, it was a good idea; the stuff would put him to sleep if the drinks he had downed didn't get him first.

The stiff did not down his drink immediately, as we all expected him to do, but went to the center of the room again and stood silent for a few moments, while everyone wondered what was coming next. After a full minute, the character raised his glass and turning to the crowd standing at the bar gave the old familiar toast: "Here's to crime, may prostitution flourish," followed by "Good-bye, friends and others, I am leaving for hell, I'll be waiting for you."

There was not a cowhand or stiff in the crowd who did not wish him a pleasant journey, not to waste time getting started, tell the devil to keep him and not send him back, and to tell the cowhands that he would meet there to keep the fires stoked until they arrived, and many other messages and wishes for a pleasant journey.

When the well-wishers' banter ceased, the character lifted his glass to his lips as if to down the drink, then thought better of the idea and recited a poem none of us had heard before, and which I have not heard again since. All I remember are parts of two of the verses: "At the close of our existence as we climb life's golden stair and the chilling winds of autumn softly waft our silvery hair. . . ." And another: "Oh how my spine would tingle on a wet and muddy day when a maiden showed her ankle in that naughty, funny way. . . ."

At the word "way" he stopped, gulped his drink, and started to go on with the poem as he fell to the floor. We left him there to wear it off, and to get up again later perhaps to finish the poem he had started. He lay there quite a time before a cowhand stepped over to wake him or get him into a chair. He stepped back, surprise and something else written on his face as he said: "Why the damned bum's dead." They called Doc LeFerge, and he confirmed the diagnosis and added, after he had looked the body over and saw burns on the lips, "He killed himself, boys, he took cyanide."

Pete and I left Dolores on Christmas day and landed in Telluride, where I spent a couple of weeks on a busman's holiday, looking over the Liberty Bell, Tom Boy, and others of the big gold producers. Pete had no desire to remain in Telluride, or any other of the mining camps for that matter, and headed for New York to marry a girl who had stalled him while she made up her mind whether to marry him or not.

While Pete was in Colorado, distance had done some good, for the girl's heart grew fonder and developed a first class ache that didn't hurt Pete. In a letter he received from her when we hit Dolores, she said that she would marry him the day he arrived in New York. So far as Pete was concerned, he had it made. He was in a Seventh Heaven and couldn't wait. All of the oxen that pulled emigrant wagons across the plains, hitched together as one team, couldn't have held him.

Pete wrote from New York that he made it in time to say "I do" as the old year went out and for her to get in the finishing "I do" as bells tolled the new. Pete's mining career was short, but while it lasted, he stood on his own and made good. He did a fine job with his marriage, too, and became an executive in a large import and export company in New York. Sweet Caporal would have done well to have learned a few lessons from Pete.

When I was satisfied there was nothing I had seen in Telluride to interest me, I moved on to Newmire, ten miles down the canyon, to see what was going on at the Primos Chemical Company, which was treating vanadium ore by a secret process. As luck would have it, a job as mill shift boss was open; the graveyard stiff whose job I took over had rendered me a service by getting hurt the night before my arrival.

There was nothing much of a secret about the process that I hadn't learned when I went off at the end of my second shift, for it was a laboratory technique modified to meet the requirements for treating large tonnages instead of grams. After I had reached bedrock, it took a few days more to discover that the company was defeating its purpose and not recovering all of the vanadium it could from the ore and that a much greater tonnage of ore could be treated daily than was going through the mill.

When I had learned all I could about the process, I had the gall to carry on experiments that I hoped would double mill production and increase the percentage of recovery of the vanadium in the ore. If I slipped up anywhere or was caught in my undertaking, all hell would break loose and I would go down the hill talking to myself. A month passed and mill production had reached thirty-five tons of solution treated on each of the three shifts, with the major part of the increase during my shift.

When I was convinced that I had the bugs out of my idea and that it would work, I went to H. Boerike, the manager, and secured permission to put the idea into operation on an around the clock basis. Jack Shell, the mill foreman, was called into Boerike's office and was given instructions to begin the new procedures the following day. H. E. (Slim) Willis, my assistant, and I were to work together on the afternoon and also the graveyard shifts; Jack would continue on the day shift, heaviest of them all, with the regular shift boss, and the afternoon shift boss would be under my instructions for the eight hours he worked.

Slim and I lost a lot of sleep, but when six weeks had passed, we reached seventy tons of solution treated on my graveyard shift. The day and afternoon shifts were only a few tons lower. The increase was about double that of the usual runs of the shifts before my plan was put into operation.

Seventy tons for the best shift of each day had been my goal, but in the afternoon of the shift following the one on which we had reached seventy, I ran into trouble and was so badly injured that I had to give up and let the doctors take me over. It had almost been undertakers, who were cheated by a split of a split-second.

While passing along a row of vats, I noticed a main driveshaft bearing smoking like a chimney. There was not time to call for the oiler; so I grabbed the nearest pail of grease and went up with it on a nearby ladder to get rid of imminent trouble and possible disaster. The grease melted into the hot bearing box in fine shape, and by the time I had gotten all there was in the pail into the box, an oiler was on the way with oil by the gallons to pour into the box.

On my way to a ladder to climb down to the millroom floor, I stepped over a quarter-drive, and a cuff of my new-that-morning overalls was caught by a set-screw on a shaft collar. The shaft was making sixty revolutions, but before I could be pulled down onto it I grabbed the rim of a vat that stood behind me and held on until there was little flesh on the palms of my hands to hold on with.

Boerike and Shell, who were working on a drag conveyor below where I was, saw what was happening and made for the drive-belt to throw it from the pulleys. Which one of them threw it neither knew, but it was in time to stop my being cut in half from the waist up, although a start in that direction had been made. The only clothing I had left on were shoes and a few ribbons. Neither while I was recovering, nor at any time since, have I cared whether the mill ever passed seventy tons on a shift or went back to the original thirty or thirty-five.

When I was able to get around again, I returned to Golden to find Ted with his hands full keeping up with his studies and trying to manage Crampton and Crampton at the same time. I couldn't go into the field, so I took over the work Ted had been doing and also worked on the clays of eastern Colorado. I had recovered from my accident at the Primos mill and was ready for the field again, but old number two-forty-one refused to melt at any temperature the furnace Ted and I had designed and were using would reach. I refused to be beaten by a mere hunk of clay, so we designed and built another furnace just to beat that one clay. The furnace worked all right, in reverse—the burning chamber lining melted before that two-forty-one and ran from the exhaust of the downdraft as lava from a volcano.

After I had redesigned the furnace and was preparing to build it, I was interrupted by a visit from Shattucky Bill. I never had an opportunity to go after two-forty-one again, but I did have the satisfaction of having Monte Butler, who was in charge of the work, include the design in his report on "Clays of Eastern Colorado" and see it published in the report by the Colorado Geological Survey.

Shattucky had left Heaton on the Dolores River early in the summer when it looked as if the strike at the coal mines located between Walsenburg and Trinidad would get out of hand; trouble had been brewing for a long time. Shattucky wanted me to go with him to see what could be done to quiet things before blood would be shed.

The coal mining industry of Colorado was dominated by the Colorado Fuel and Iron Company, the C. F. and I., a Rockefeller controlled outfit, whose officials from pit bosses to the top had little or no sympathy and little consideration for working stiffs. What the stiffs were paid didn't give them a chance; living and working conditions were a disgrace to the operating companies, which worked the mines on the theory that "men are cheaper than timber."

Many of the stiffs on strike were bohunks, imported by the companies employing them or furnished, at a fee, by professional importers of cheap labor. Once on a job, the bohunks had little choice other than to remain and accept whatever wage was offered, and to work under mining conditions that were a threat to their life every time they went underground.

The companies not only paid whatever they pleased, but also rented shacks to the stiffs, or put them in bunkhouses unfit for human habitation. The stiffs were obliged to trade at company stores where they were charged unreasonably high prices because there were no other stores in competition with the company-owned outfits where credit was freely given and the bill deducted from the stiff's time. Whenever a stiff complained, he was more often than not given his time, removed from his quarters, and denied the "privilege" of making purchases at the company store. The threat of deportation was ever present.

Not a few times conditions became so intolerable that a strike was called, with little or no success. The stiffs, who had no money when they went out, could not last very long and usually capitulated with gains that meant nothing. The strikes in the Walsenburg to Trinidad coal camps were a result of grievances against conditions that could no longer be tolerated. Feeling ran so high against the companies that Shattucky and I knew that the job we had come to do was an almost impossible one.

Not only were the working stiffs fighting against the coal companies, but also against officials of Colorado, who from the lowest constable on up to the governor's office, were, almost to a man, against the strikers and on the side of the coal mining companies. Particularly were they partial to the C. F. and I.

Shattucky and I found the Colorado National Guard keeping

"order," not impartial order, but the kind of order the mine operators and the C. F. and I. demanded, at every mine or town we visited from Walsenburg to Trinidad. Unrest was at its worst around Ludlow, where the strikers had established a tent colony for themselves and their families.

At Ludlow, where Shattucky and I spent most of our time, Captain Carson commanded Cavalry Troop A, and Lieutenant Linderfelt Infantry Company B, although Major Hamrock should have been in command of B Company. Food among the strikers was scarce, and when shipments sent in by the miners' unions arrived by rail, Troop A and Company B, which were camped near the depot, would confiscate them on the pretext that arms and ammunition were concealed in the food cases.

During January, February and March, Horace Hawkins, attorney for the United Mine Workers, E. L. Doyle, a district secretary, Louis Tikis, and a score of others did a real job in smoothing things out. It appeared that, if the National Guard were withdrawn, a peaceful settlement might be made. A short-lived peace came when Governor Ammons recalled the National Guard and left the mine operators and the C. F. and I, to do as they pleased.

Company B under Lieutenant Linderfelt was left in Ludlow with thirty to forty "guardsmen," mostly mine guards, strikebreakers, and lesser company officials. After the National Guard pulled out, the operators of almost every coal mining town recruited companies of "militia," which were designated as "National Guard." The musters of the militia companies read as if they were payrolls of the mine operators. To the disgrace of the State of Colorado the militia companies were armed and equipped by the State, but paid by the operators, and the C. F. I., as later investigation proved.

Conditions became intolerable, and when rumors spread that the Ludlow tent colony was to be wiped out, resentment reached fever pitch. How true the rumors were was not known, but there was a precedent that the strikers had to guide them—that of the Forbes tent colony. On Sunday, April 19th, Shattucky, Louis Tikis, and others of the more level-headed, and I did all we could to ease the tension and to convince the strikers that the rumors were false. Quiet had been restored and everything was under control by late Sunday evening.

Tikis, Shattucky, and I went to see Linderfelt. We told him that if the C. F. and I. militia would cease their activities, we could guarantee that the strikers would cause no trouble. Linderfelt told us to go to hell, that he was in command and would be the judge of how to handle the strikers. Tikis, a cool-headed, intelligent man pleaded with Linderfelt to no avail. When we returned to our quarters, we were convinced that Linderfelt would do nothing to prevent serious trouble.

We were not mistaken. Early in the morning of the 20th, Company B, under command of Linderfelt, occupied the hill overlooking the tent colony. The hill gave the militia command not only of the colony but also of the surrounding hills and roads that led to it. A machine gun was in view on a rise on the hill, and the C. F. and I. militia stood with rifles ready and appeared to be awaiting orders to attack. The strikers hastily armed themselves as best they could with the few defensive weapons at their disposal and waited.

Shattucky, Tikis, and I, with a score of others did our best to control the strikers. The strikers did not want trouble, but were preparing to meet it. There might have been none had not Linderfelt's men let go with a couple of bombs, followed immediately by machine-gun and rifle fire.

Company B did not concentrate its fire on the strikers, most of whom had left the tents and had taken positions in the creek beds and on nearby hills, but poured rifle and machine gun fire into the strikers' tent colony where there were defenseless women and children.

The fighting lasted from morning until after dark, and during that time the militia swarmed into the colony, looted it of everything the men could lay their hands on, and attacked the women and children. The reign of terror finally ended when the militiamen poured coal oil on the tents and burned the strikers' shelters to the ground.

Shattucky and I got away, but Tikis and two others were captured, disarmed, and stripped of everything but outer garments and taken to Lieutenant Linderfelt's headquarters. We did not know it at the time, but it was disclosed in investigations that were made later, that Linderfelt, in a rage, brought a rifle butt down on Tikis' head, breaking the rifle stock and killing Tikis. Not to be denied their fun, his aids completed the job by killing the two other captives and putting bullets into the already dead Tikis.

Shattucky and I helped the strikers dig seven children and three women from the wreckage of their burned tent house homes. There were in all twenty-one killed in the massacre, and before order was restored thirty strikers were known to have died by militia gunfire. One militiaman was killed. While we were occupied in recovering the bodies, Lieutenant Linderfelt and some of his militiamen came to where we were working and took us prisoners. With others who had been with us, we were searched and everything on us taken away; money and personal effects were confiscated with whatever else was found. After the search and seizure, we were escorted out of town and told that if we returned we would be dealt with.

Every miner, coal and hard-rock, in Colorado was in arms and a statewide uprising was threatened with thousands volunteering to go to the aid of the strikers and their famiiles. It was a situation that the C. F. and I. and the State of Colorado had not anticipated and could not control. Governor Ammons panicked and called on President Wilson for federal troops to put down the "insurrection" against the State of Colorado.

Before the federal troops arrived, the strikers took control of every coal mine between Ludlow and Trinidad and burned the buildings around them. Meanwhile, Governor Ammons called out the legitimate National Guard on April 23rd, and General Chase, who commanded it, occupied the mining areas between Walsenburg and Ludlow. The strikers were in such force south of Ludlow that the National Guard did not go farther than there; to have done so might have precipitated trouble more serious than that which had already occurred.

To make the situation more difficult, many guardsmen were hard-rock miners, and not a few were from the northern coal mines. They knew what the strike was about and what had happened when illegitimate militia were organized by the mine operators, supported by the State of Colorado, and erroneously called "National Guard." To complicate the difficulties, General Chase had made agreements with the strikers through Hawkins and Doyle, and then had violated them. The strikers no longer had confiidence in the officials of Colorado, nor in the officers of the National Guard. Had the Federal troops not arrived when they did, more serious trouble would have developed.

Shattucky and I were in Trinidad when the Federal troops

took over the whole area. Since there was nothing more that we could do to help improve the situation, we returned to Denver. I left Shattucky in Hawkins' office with Doyle and other labor leaders. I never saw him again. I went to Golden to get the stink of things I had seen out of my system and to forget: the stink departed, but the memory has not.

Because of the coal miners' strike and the resulting labor unrest, the fear that trouble might spread to the gold and silver mines throughout Colorado prevented activity among the small mine operators that Crampton and Crampton depended upon for its business. As a result of the mine operators' marking time, business slacked up. I was considering going back to Arizona or California where I thought business would be better, when a letter arrived from Pete Buol asking me if I would come to Las Vegas and do some work for him.

I replied and told Pete that I would be in Las Vegas as soon as possible and then waited until Ted was graduated with high honors and a Phi Beta Kappa key. After graduating, Ted went to the Mountain Chief mine near Quartzburg, in the Boise Basin of Idaho. Crampton and Crampton had been doing some work for the operating company that I had been attending to for A. C. Gallup, its president, and Bill Fairchild, the treasurer. The Mountain Chief was the first chore Ted did as a full-fledged mining and metallurgical engineer with his E. M. degree from Mines.

Goodsprings, Nevada, probably 1916 or 1917. Office of Jenson and Crampton with Holt caterpillar and trailers loaded with ore on way to Jean, the shipping point for ore. Crampton Collection.

Carbonate King Mine, a Crampton and Crampton mine operation from 1915 to 1920, near Roach, Nevada. Burro train packing ore from the mine to loading platform from which it was hauled to Roach for railroad shipment. Crampton Collection.

Piute Mine, Cima, California, spring, 1915. The Piute was a Crampton and Crampton mine operation from 1915 to 1920. From left: Sully going into hoist house, the author standing under gallows frame at collar of shaft, John T facing camera. Crampton Collection.

Piute Mine, Cima, California, June, 1915, a few weeks before Sully was killed. From left, John T, Sully, Bob English, Jack Coyle, Steven Swen, Hal Wright. Crampton Collection.

Piute Mine, Cima, California, July 11, 1915. Pete (Senator Buol) stands beside Locomobile while the author prepares to crank the engine, an hour after Sully was buried. Crampton Collection.

SULLY USES A METAL CLEANING SPOON

Soon after Ted was graduated from Mines, John T and I left for Las Vegas to see Pete Buol. Pete told me that he had taken over the Boss platinum mine near Goodsprings, fifty miles west of Las Vegas. The Boss was a freak, for its ores contained platinum and palladium, in addition to gold, silver, copper, and lead in amounts large enough to be commercial. Some of the high-grade ore ran as much as $50,000 a ton. The mine had attracted a great deal of attention among mining engineers because there was no other like it in the United States. Because platinum-palladium was in such quantities in each ton of ore, occasionally as much as several thousand ounces and frequently several hundred, the United States government was interested. Platinum was a strategic war mineral, and Adolph Knopf of the United States Geological Survey came to the mine to make a report on it for the government.

Pete was interested in the Boss with W. C. Price, a former stagecoach driver on the Elko-Tuscarora run, who, I learned later, knew Ed Devenney and had staged with him. Price had quit driving stages when he made money on some wildcat oil stock that he had purchased, and wanted to get into mining. Pete sold him on the idea that the Boss was the mine he wanted. Price's brother of "Who's Your Tailor?" fame, a mail-order "made-to-your-measure" outfit of Chicago, was also interested.

The ore from the Boss had such a complexity of minerals that smelters to which it was sold would not pay for all of them. Platinum and palladium was paid for at the rate of twenty dollars an ounce, yet they were worth more than double that amount. Often platinum and palladium were not paid for at all.

My job was to develop a metallurgical process for the recovery of each mineral separately, in a mill to be built at the mine. When I worked out the process, we found that the cost of treatment exceeded the value of the minerals in the average ore. Only the high grade could be treated profitably. After a year had passed, Pete and the Prices let the mine go to someone who wanted to mine for platinum and to worry about getting paid for it.

The first problem was to develop an assay method whereby the value of each of the metals in the ore could be determined within twelve hours after the ore had been mined. Before such a method was found, ore from each face was put in a separate bin to remain there until its value was known, then to be thrown over the dump as waste, shipped to a smelter, or sent over the ore dump to remain until an economical metallurgical process could be developed to treat it profitably.

Getting an assaying method was not difficult, but other trouble made up for what was missing on the technical part. The assay office was a long, high-roofed, four-room building, with unscreened windows. To heat from the assay furnaces there was added that from the red-painted corrugated iron roof. The interior of the assay office was a perfect place for the devil to experiment on tortures for mining engineers who slipped from the straight and narrow, and for a time he took full advantage of the facilities that the office afforded for his devilish experiments.

At first I fought the devil with ammonia and sulphur fumes. I should have known better. The devil uses vast quantities of sulphur in hell, and ammonia probably gave him new ideas. Failing in the use of ammonia and sulphur, I tried sulphuric acid and cyanide, which probably pleased the old fiend, because if I bungled in the use of it he would get me the sooner. When I found that I could not beat the devil at his own game, I accepted his experimental punishments as best I could.

A herd of prospectors' burros, tame and docile from long association with unpredictable masters, made use of the shady north side of the assay office for their afternoon siestas. With the burros were myriads of flies, common, horse, bot, buffalo, buzz, and other flies of species unseen before.

As they rested the burros unpeacefully swung their tails and rubbed their bodies against the building to brush or rub the flies away. As a result of the burros' activities the building shook, and flies swarmed through the unscreened windows. The torture never came when I was occupied with some easily performed chore, but always when I was doing something that required the utmost care and precision.

The job was difficult enough with the building shaking, but the flies doubled their persistence when I poured white molten slag from a crucible into one of the iron molds. Sweat bubbled from my arms and face in the oven-like building, and as I poured slag into a mold, alighting flies would loosen beads of sweat, which would drop into the mold, with a resulting explosion that threw the white hot material in every direction and often set the walls and floor afire. My face was burned and scarred so badly that I seemed to have a bad case of chicken pox.

One afternoon I had an inspiration. I took a grease-gun from my Model T, filled it with turpentine, and gave all but one of the burros a shot of it, in the rear. I got away from kicking hoofs and twisting bodies through a window from which I watched the madly kicking and twisting burros head for the wood-ashes pile at the cookhouse and roll in it until the dust soon obscured vision of them. Soon one after the other of the burros emerged from the cloud and headed for the valley. They never returned. The devil was defeated by a simple use of turpentine.

About the time I was getting under way on Pete's problems, the war in Europe was going strong, and as a result the Goodsprings and Yellow Pine districts boomed because of their large zinc and lead deposits. I sent for Ted, and we established the Goodsprings Assay Office where we conducted a mining engineering, metallurgical, and assaying business. John T and Sully were already at the Boss and took charge of the mines and prospects we operated and assisted in mine examinations.

The business of Crampton and Crampton expanded rapidly, and we took J. C. Jensen into the firm. Jensen had an ore sampling plant at Jean, eight miles from Goodsprings, where the Los Angeles and Salt Lake Railroad had a siding and freight loading platform. Ores from the Goodsprings and Yellow Pine districts were shipped to smelters from Jean. Jensen and Crampton purchased ores and paid more for them than the sellers could ordinarily get because of the favorable purchase and sale contracts Jensen and Crampton had been able to negotiate for ore purchases by the smelters.

Crampton and Crampton extended its activities to include the operation of zinc and lead ore prospects and small mines. The business grew fast, and we brought in two Holt tractors and fifty ore hauling trailers. Kenny Nikrent, one of the Nikrent brothers who had driven the Buick that had won the Los Angeles-Phoenix road race, handled the Holt tractor situation for us. Kenney sold me the Buick he and his brother had driven, and often borrowed it when there were races held on one of the nearby dry lakes. Kenney won all but a few of the races in which he entered the Buick.

The Goodsprings Assay Office did a landoffice business assaying ore samples for prospectors, but did not take in enough money from them to pay the cost of the work. We charged prospectors, desert rats, and any of the old timers fifteen cents for gold, silver, lead, or zinc, two bits for any two minerals, and four bits for any combination of four; all others were charged the standard and prevailing rates for assaying.

No prospector who sent samples to be assayed was billed nor were any asked to pay. There was not a prospector or desert rat who did not hear how soft we were. The results were astonishing. Samples by the thousands poured into the assay office from hundreds of prospectors, and as a result Crampton and Crampton had first, and first-hand, information about every outcrop and prospect hole within a hundred miles.

If a prospector's assays looked all right, Sully, Ted, or John T went out to take a look at the claims. If they looked good, a lease with an option to purchase was taken and work started as soon as a crew could be placed on the job. Sometimes claims would be purchased; the Piute and Green Monster Extensions were two of these and paid off their purchase price many times over.

John T, Sully, and Ted were the best ever when it came to sizing up prospects. They didn't make many misses on the good looking ones, and even when they took a lease on a prospect on a hunch, the long chances made good more times than not. One of the long chances was the Carbonate King, a zinc mine just over the state line in California, which ran the Piute, an almost sure thing from the time the first pick was put into the ground, a close second.

Sully had elected to take a look at the Piute when it first came to our attention. Sully liked it and said that he could make a mine of it if given the chance. I turned the job over to him and told him he wouldn't get paid if the Piute didn't come through. Sully never missed a payday.

The Piute was doing so well and development opening up new ore so often that John T or I would often make the sixty-mile trip from Goodsprings to it to see what was going on. One morning Pete, who was in Goodsprings on Boss Mine business, asked John T and me if one of us would take him to the Piute. He had heard so much about the mine that he wanted to see if it was as good as he had heard that it was.

Pete had no illusions about mines and mine owners. He had to be shown. Pete's definition of a mine, "A hole in the ground owned by a liar," was the same as that of Mark Twain, and Pete wanted to know if John T and I were or were not liars. John T and I went with him, on one of the infrequent trips we took together to the mine at the same time.

We made the Piute an hour before lunch, and the shift was in the hole. So was Sully. The hoistman let us down to the hundred-fifty, where he thought we would find Sully. We looked around and finally came to where he was working up a raise in a stope, cleaning out a missed hole and getting it ready to reload.

Sully went down the raise with us and took us to the two hundred, where a two-foot-wide high-grade ore shoot had been broken into the day before. Just before tally, we went up the shaft, and Sully went back to his cleaning-reloading chore. Sully wanted to get the shot fired so that the face could be drilled and shot on the afternoon half of the shift.

We were at the table eating lunch when we felt the ground shake and heard a dull boom underground. Sully had shot the hole and would soon be up. We went on eating, but when Sully did not show up, John T and I left the table, went to Sully's tent, found no Sully, and then to the collar of the shaft. There was still no sign of Sully nor any sound of his coming up the ladderway, so we went down.

Sully was in the stope, the pointed end of a metal cleaning spoon driven deep into his left chest, and his face torn by rock from the blasted hole. John T went for a piece of lagging, and when he returned with it, we fastened Sully to it and lowered him through the raise to the level below. We trammed him to the station and rang the accident bell signal. We waited a few minutes before we heard the hoist start, followed by sounds of men coming down the ladderway.

When we heard the hoist we signalled for the bucket. After it stopped at the station, we lifted Sully into it and put our arms around him to hold him tight to the cable so that he would not fall out. We signalled to hoist and were on our way up the shaft before the stiffs on the way down the ladderway reached the station.

We would not let any of the stiffs help take Sully from the bucket, nor help carry him to his tent house. We wanted to be alone with him until we laid him in his resting place. Pete, who understood the ties that bound Sully, John T, and me together, kept the stiffs away and left us alone.

John T and I had not spoken from the time we left the cook house to go after Sully. When we had laid him on his bed, we sat with him, one on either side. We once in a while said something to Sully, while tears flowed like rivers. While we were with Sully, Pete had the stiffs dig a grave on a little rise not far from camp, and make a coffin from the best mine lagging they could find.

When we left the tent, Pete was sitting outside, and the other stiffs not far from him. Not a few had tears that they could not stop. The others held the tears back. Pete got to his feet and motioned to the coffin he and the stiffs had made. John T and I carried it into the tent house, carefully wrapped Sully's bedroll around him, and laid him in it; then we closed the coffin and nailed down the lid. We knelt beside Sully and asked the Lord to be good to him. After that we went to the door, and the stiffs came in and carried him to his grave.

Before Sully was lowered, Peter read the Twenty-third Psalm, and we all said the Lord's Prayer. As the earth was mucked into Sully's grave, over rocks that would keep coyotes out of it, the trio that had left Chicago together more than ten years earlier was no more.

Pete drove the Model T back to the Boss, where we found C. W. and Nevius making themselves at home while they waited for us to return. C. W. and Nevius had left New York together on a trip to the Panama Pacific International Exposition, had stopped at the Chloride mine, and then headed for the Boss to pick me up and take me with them.

Nevius wanted to reach San Francisco and the Panama-Pacific International Exposition as soon as he could. He was one of the judges of the mineral collections on display at the fair, and responsible for the final determination of winning exhibits. Nevius wanted as much time as he could get to study the minerals in each exhibit because of the technicalities involved.

After spending a few days at the Boss, which neither C. W. nor Nevius had seen, we headed for San Francisco in my Buick. C. W. and Nevius asked John T to go with us, and he did. I would not have gone without him.

C. W., John T, and I took a look at the exposition and then drove to the Panama-California Exposition at San Diego, which was running in competition with the exposition in San Francisco. When that was looked over, we retraced our route to San Francisco, stopping over at the quicksilver mine, north of Jolon, to see what was doing there. The mine was closed down. The new owners had taken out the best ore, left the low grade that could not be mined profitably, and departed. Locke had been wise to tell his London backers to sell.

From any point I looked at it, the Exposition was a success. Nevius awarded me the gold medal for the best exhibit of platinum ores; there were no others, so the honor was an empty one. Another gold medal award I received was a different matter. There were other entries besides mine, but because mine contained an unusual crystal specimen of a copper-antimony-arsenic compound from the Wren mine, Nevius made the award to me.

After the awards were made, Nevius brought T. A. Rickard, editor of the *Mining and Scientific Press,* over to introduce him to John T, C. W., and me. Rickard wanted me to prepare a series of articles on the Boss Mine, its geology, ore occurrence, and the metallurgical problems that had to be met in order to recover the minerals in its complex ores. Rickard's request was a great compliment, and I agreed to prepare the articles for publication in the M. and S. P. Unfortunately, the pressures of other work prevented my preparing more than two of several articles I had agreed to write for him.

Rickard was a mining engineer of great ability, and he exhibited an understanding of the problems confronting the mining industry and the men engaged in it. Rickard's ability and editorial policy had made the M. and S. P. the most important and reliable of the periodicals in the West that were dedicated to mining and that served engineers who were concerned with the practical aspects of mining.

The *Mining and Scientific Press* was eventually absorbed by the *Engineering and Mining Journal,* which was published in New York and had an editorial policy that was world wide in scope and was not as concerned with western mining as the M. and S. P. had been. The *Engineering and Mining Journal,* in addition to its general mining coverage, was a valuable periodical to the mining industry because of its dependability in metal market quotations which were accepted as authoritative. It did not, however, take the place of the M. and S. P., which had been a publication devoted for the most part to western mining.

The pressing demand for a dependable and reliable mining periodical that would serve the western mining states was filled by the *Mining Journal,* a western publication owned, published, and edited in Phoenix by Charles F. Willis. Willis not only adhered to and followed Rickard's editorial policy, but also added something that made of his *Mining Journal* a greater publication than had been the *Mining and Scientific Press.*

Willis was aware of and understood the problems of the small mine owner, and the difficulties he had in interesting capital in small and undeveloped properties, even when they had every indication of someday becoming a mine. The *Mining Journal,* under the editorship of Willis, achieved results for the owner of small mines that no publication had achieved before.

When Willis "retired" and sold the *Mining Journal* to the publishers of the *Mining World,* he could not remain inactive so long as the small mine owner needed assistance. Willis came out of retirement and became secretary of the Arizona Small Mine Operators Association in order that he could continue to give help to the men who were the backbone of the mining industry. Fortunately for the industry, in 1950, the *Mining World* publisher engaged George O. Argall, Jr., a Colorado Mines graduate with vision and understanding, as its editor, and the policy of Rickard and Willis has been continued since that time.

The day after the Exposition awards were made, we drove to Nevada City. Nevius wanted to try again to buy the Murchie Mine, which he had examined earlier, when I was his assistant. When Nevius made his first attempt, battling owners prevented his purchasing it. The situation had not changed, and he also failed on his second try. Some years later the Newmont Mining Company bought the mine and did well with it.

After failing to purchase the Murchie, Nevius was ready to leave the Nevada City-Grass Valley area and return to San Francisco, and C. W. wanted to get back to New York. We remained long enough, however, for Nevius and C. W. to visit other mining engineers whom they knew in this area. Among those whom Nevius or C. W. introduced me to were George Star, Arthur Foote, and Mike Brock, who more than fifteen years later, after I had returned from a long stay in China, served with me on a committee appointed by Governor Rolph, in 1932, to investigate mining and workmen's compensation insurance problems in California. The association on the committee lasted until 1934 and substantial results were achieved; and lifetime friendships were established.

I also met James D. Stewart, of Auburn; W. W. Waggoner, and Bill McGuire, both of Nevada City, placer mine operators; and E. G. Kinyon, editor of the *Morning Union* published in Grass Valley. Jim Stewart and I worked together from early 1932 on a bill Jim had had Senator Jerry Sewell of Roseville introduce in the California Legislature. The bill, when passed, would make it possible to resume hydraulic placer mining operations in California, with protection of the farmers and other water users who had suffered serious damages in earlier years, and it also protected mine operators. Jim's bill, when it became law, would prevent the old rough-shod, to-hell-with-the-farmer methods of the past.

At the time Jim was working to have his bill passed, I was the manager of the Mining Department of the Los Angeles Chamber of Commerce. A. G. Arnoll, whom I knew well, was the Chamber's secretary and manager. I was also, and had been for some years, advisor to A. G. on technical and engineering matters that affected the Los Angeles area, particularly on its water and industrial problems.

Jim and I convinced A. G. and the Los Angeles Chamber of Commerce that resumption of hydraulic placer mining would benefit southern California and particularly Los Angeles manufacturing firms, and the Mining Committee of the Chamber was instructed to get behind Jim's bill. The Los Angeles *Times* and the Los Angeles *Examiner* also helped by giving placer mining considerable publicity.

Jim fought almost single handedly for over twenty years to make it possible for hydraulic placer mining to be resumed and had spent his time and many thousands of dollars of his own money freely to attain that result. Jim's sincerity, integrity, unselfish motives, and perserverance finally resulted in success for the hydraulic placer mining industry.

Jim put on a strong and effective campaign in northern California, and with the help that the Los Angeles Chamber was able to give, Jim's bill passed both houses of the legislature by good majorities. When Jim eventually succeeded where all others had failed, other hydraulic placer mine operators did not take advantage of his successful endeavors and hydraulic placer mining continued to remain dormant.

Above: *The Goodsprings, Nevada, flood, August 21, 1916.* Below: *Art Clark, left, and Charlie Waring. The red-light "line," surrounded by water, in the background.* Crampton Collection.

Piute Mine, Cima, California, spring, 1917. Shorty Harris with his burros on one of his grubstaked prospecting trips, during which he made a visit to the author. Crampton Collection.

Kolcheck Mine, near Ely, Nevada, summer, 1917. From left: Shoshone Indian, Alec Kolcheck, a miner, Judge Dan MacDonald. Crampton Collection.

Green Monster Extension, Goodsprings, Nevada. Standing by a prospect shaft, from left, are Sully, Link Shaw, Charlie Waring, an unidentified man, and the author. Crampton Collection.

West of Paranagat Lake, Nevada, on road between Pioche and Las Vegas. A road sign typical of those on the desert as late as 1920. Photo by Pete (Senator Buol) while on a trip with the author in 1917. Crampton Collection.

Little Smokey Valley, Nevada, spring, 1920. The author making repairs to broken-down Model T while on an exploration trip with Pete Buol. Crampton Collection.

Approaching the Fairview pass summit on the road from Sunnyside to Pioche, Nevada, with Pete Buol driving and churning snow in his ill-fated Studebaker. The car broke down, for the third and last time, two hundred yards beyond where this picture was taken. Winter, 1921. Crampton Collection.

Within one hundred yards of Fairview pass summit, on Sunnyside-Pioche road, Pete Buol's Studebaker broke down for the last time. The covered buckboard is that of the sheriff and his posse from Ely. Within the hour following, posses from Pioche and Las Vegas arrived to "rescue" the Senator. Winter, 1921. Crampton Collection.

A group at Atlanta, Nevada, where C. W. Wheelock was operating a silver prospect, winter, 1921. Standing, left to right: Jack Fitzgerald, "C. W." Wheelock, and Vic Husen; Sam Waring, seated. Crampton Collection.

Pioche, Nevada, summer, 1921. A reunion of a few of the old timers and some not so old. From left: J. Nelson Nevius, Earl Godbe, Ruth Waring, Jack Fitzgerald, Doc (J. E.) Mayers, Jimmy O'Brien, C. C. Waring. Crampton Collection.

THREE POSSES FOR THE SENATOR

When Nevius and C. W. had neither business nor more mining friends to keep them in the Nevada City-Grass Valley area, John T and I saw Nevius board a westbound train and C. W. an eastbound one at Colfax. We ourselves headed for Goodsprings and the Boss Mine. Awaiting me at the Boss was an urgent note from Pete asking me to join him in Las Vegas as quickly as I could after my return from the north. In his letter Pete told me that he wanted me with him when he went to examine a silver prospect near Tecopa and another adjoining the Cerro Gordo silver mine in the mountain range east of Keeler.

Pete was continually making trips to some prospect hole, and whenever I was free he asked me to go with him. He was always in a hurry to get started, but it didn't matter when we returned. Pete's trips were never dull nor were they uneventful. He invariably took the longest route to reach a destination unless the roads were bad over the shortest one, and then he would take that. Whether a trip with him was long or short, it would be one to remember. Always we had fun, but also more often than not, we got into some kind of trouble.

I drove my Model T to Las Vegas, and Pete and I took the trip in it. Pete's cars were always breaking down, and I had no faith in anything he ever drove. I was certain the Model T would get us where we were going and back again. There was nothing at Tecopa, so we went on to Owens Lake and Keeler, east of it, stopped at the Cerro Gordo, and then went over the hill about a mile to the prospect Pete wanted to look at.

The prospect had a three hundred foot tunnel with several

raises that had started up on ore that petered out before many rounds had been shot. We had no chance to see whether there was ore in the only winze, but Pete found that there was water. Carelessness led Pete into it, for he was holding his carbide lamp near the roof and not looking where he was going when he disappeared into the waterfilled winze. I got Pete out without any difficulty, but I had to listen to his profanity while he reviled the last hard-rock stiff who had been in the tunnel for not putting the floor boards over the winze.

On the return trip Pete insisted on taking one of the "worst road" routes, over almost unused "marks" that went through Silver Dry Lake, then crossed the desert washes and passed to Kingston and from there to Goodsprings. It was bad going, and when we reached a short sandy slope, the Model T carbureter coughed a few times and gave up. The carburetor was not getting gasoline, and we assumed our gas tank was empty. When we walked to the top of the slope we saw a lake, square miles of it. There was no lake shown on any map of the area nor did we know of any.

The only explanation that Pete or I could think of was that in night driving we had turned off the trail, and were lost. After we looked the situation over, we saw smoke from something that might be a house, two miles from us and across the lake. It was seven miles around to the house, which turned out to be the town of Silver Lake. Silver Dry Lake had flooded in a recent storm.

We retraced our around-the-lake route in a Model T that we rented with its owner going along with a five-gallon can of gasoline to put in the gas tank. When we reached the Model T, the owner-driver took a look at the situation and laughed. He got out of his Model T, walked to ours and gave it a push backwards until it reached the bottom of the slope. Then he cranked our Model T, started it, turned it around, backed up and over the grade. We had not run out of gasoline, but the gasoline in the half-full tank under the driver's seat wouldn't run up hill and into the carburertor. When turned around and driven backwards with the gas tank on the up-hill side, the gasoline flowed to the carburetor all right. We made Las Vegas without further incidents.

Another trip I took with Pete in October and November, 1919, into the same desert area was one to look at a deposit of alunite, an ore rich in aluminum, on the west slope of the Funeral Mountains in the Amargosa Desert. As usual there was nothing that

Pete thought he wanted, and we went on to look at a manganese prospect Pete had heard about near Owl Hole Spring at the south end of Death Valley. Pete insisted on taking the longer and most dangerous route, to Furnace Creek Ranch. From the ranch we would go south through Death Valley to Denning, an old and almost abandoned mining camp, then west to the Owl Hole prospect.

It was fall and I didn't protest too much, for duck shooting on some of the lakes in the lower part of Death Valley would be good. It had not been long since Shorty Harris and I had found Old John Lamoigne that Pete and I arrived at the Furnace Creek ranch where we intended to camp before going south in the valley the following morning. We made camp by the irrigation ditch and its warm soft water, and then drifted up to the ranchhouse to have a visit with Oscar Denton, who managed it for the Pacific Coast Borax Company.

The ranch already had a visitor. He was camped in a tent with a canvas fly over it, and water was fed from a pipe line that had been hooked on to the windmill pump especially for the purpose of dripping water on the tent quarters. Inside the air was as cool as our "desert ice box" coolers, the only difference being that water, here, was fed to it through a pipe instead of dripping from a pan. Inside the tent, established as comfortably as if in a swank hotel, was Zane Grey.

Grey had drifted to Furnace Creek for local color for a story of Death Valley he intended to write for one of the large circulation, nationally-distributed weeklies. One could never have described Pete as "indirect," and he told Zane Grey that, if he really wanted local color, he should pull down his tent and throw a bedroll where we were and camp with us. Grey told us he couldn't write or think if he were uncomfortable, and we let it go at that.

The next morning while Pete was making breakfast, Grey stopped by our camp and told us he was going into the valley to get some pictures, and pointed out the place he intended to go. He carried equipment that he should have had a burro to pack for him—a camera tripod, cameras, cases holding we knew not what, a six-gun, and a pint-size canteen of water.

Pete told Grey that he better get one of the Valley Indian ranchhands to go with him to help carry the load and to tell him where to go and where not to go. He also told him that if he didn't take someone along who knew the valley, he was in for trouble. Grey as much as told Pete that he didn't know what he was talking about, but Grey found out that he was mistaken about Pete not long after he started into the valley alone.

There was no doubt in our minds that Grey was walking into trouble. To prepare for what we were certain would soon happen, we went to the ranchhouse, talked to Oscar, and had him round up two of his Panamint Indian ranchhands. Then we returned to our camp by the irrigation ditch, unloaded the Model T of its load, weighted the rear end with rocks, packed a few pieces of necessary equipment, and waited. We did not have to wait long.

A cloud of dust began to rise in the still cool air, and we were off in its direction in the Model T with the foreman, two Indians, Pete, and I. As we drove toward the dust cloud, it became larger and larger, and when we reached it, we could not see what was within it, but we knew it was Grey.

It took but a short time to take two boards from the Model T, work out on them to begin to extricate Grey from the mess he was in. Grey had walked over a bed of soft alkali, fine-powdered and dry, whose crust had supported him for fifty feet. As he approached the center of the dry alkali "pond," the crust broke, and down he went into it. The load he carried on him made his weight too much for the crust to hold up.

The harder that Grey had fought to get out, the more crust he had broken. Also the weight of the equipment that he had strapped to him made it impossible for him not to break the crust as he fought to get up on it.

Back at the ranch Grey had some of the Indians take his tent down and pack it in his car, together with his other equipment. He was out of Death Valley before dark, and we learned later that he made Death Valley Junction, in the Amargosa Desert, in time to wash up and spend the night.

The following day Pete and I made the manganese prospect near Owl Hole Springs. It was a fine strong outcrop and a good prospect, but the ore wouldn't be worth mining until there was another war and the price of the metal went up. We made Las Vegas without difficulty and without any more adventures or misadventures.

The last time I went with Pete on one of his wild trips to look at prospects was in the winter of 1921, shortly before I planned to leave for China. Pete had two prospects that were something special, and he wanted me along to look at them with him before I went away and he couldn't have me again to help bear his burdens or share his self-imposed troubles.

One of the prospects was in the Quinn Canyon Mountains east of Adaven, the other east of Sunnyside in the Ely Mountains north of Pioche. There was nothing about the trip that would be easy even in good weather, which it was not, but with any kind of luck we should not be gone from Las Vegas more than five days. Pete had to be in Carson City, the State Capitol, in a week to attend some very important political chores.

A storm was threatening, the nights were bitter cold, and the roads a muddy mess. I wanted to make the trip in my Model T, which I was certain would return us to Las Vegas by the time Pete had to be back there. Pete, however, wanted to go in his Studebaker. The Studebaker was continually tearing its rear-end transmission to pieces under the rough treatment the car got on the same kind of desert and mountain roads that Pete and I would have to travel to the prospects and return.

I had visions of walking fifty or sixty miles for help if the Studebaker did break down, with resulting delays in getting Pete back to Las Vegas in time to get to Carson City for his political commitments, and I continued to insist on going in my Model T. Pete would have none of my Model T, and I capitulated only when he agreed to take two rear-end and transmission assemblies with us, so that we could repair the car without walking for help if it did break down.

We pulled out of Las Vegas before the sun rose and made Hiko at almost dark. We had planned to be in Adaven, fifty miles farther on. The drive had been a hard one all of the way and especially across some of the low stretches that had once been dry lake beds. The lower Paranagat Lake gave us trouble, too; there was more water in it than I had ever seen before, and it covered the road. We had to abandon the road while we manipulated our way around the lake until we finally again reached the road to Hiko on the other side.

A start was made for Adaven before daylight so that we could cover the thirty miles across Coal Valley dry lake before the sun rose and softened the ice crust that formed on it during bitterly cold nights. Half way across the dry lake the wheels went onto a thin spot in the ice-frosted surface and the Studebaker sank into the soft mud underneath. I cut brush to put under the wheels to hold the car from going deeper into the mud. As Pete tried to ease it over the mass of stuff I had cut and onto more solid ground, the rear end gears tore themselves to pieces.

It took the remainder of that day and almost all of the following to dig enough mud from under the car to hold it in a position that gave us a chance to work, and to replace the broken assembly with one of the spares we had brought along. With the aid of more brush Pete worked the car from the mud hole, and we reached the shore of the lake as darkness closed in on us and a hard, wind-blown snow storm suddenly let go.

The blinding snow, aided and abbetted by darkness, made vision of the road almost impossible. I wanted to stop and sleep in the car all night, as we had on the dry lake when we couldn't work on the car because of darkness and cold. Pete insisted we go on to Adaven, about ten miles from the shore of the dry lake. We had traveled five of the ten miles when we went off the road in the poor visibility, and in trying to get it back we ran into a shallow sandy wash. While Pete was fighting the snow and slippery sand, the wheels spun a few times and again the rear end tore itself to pieces.

We walked the five miles to Adaven, shacked up with a prospector living there, and in the morning started early to bring the Studebaker into camp. Fortunately for us, we didn't have to repair the car again in the open. The snow would have been almost as bad as the mud of the dry lake. There was a stake body truck in camp, and its owner offered to tow the broken down car into Adaven for repairs.

The remainder of the day and until mid-afternoon of the day following were used in getting the one remaining spare rear end assembly in place so the car could run again. When repairs had been completed, Pete took a look at the prospect we had come to see, and by the time we returned to camp it was almost dark and too late to go on; so we remained in Adaven another night.

The storm had not lasted, and there was not enough snow on the ground to interfere with driving. We were five days out from Las Vegas, and Pete was going to be late for his Carson City appointment if we went on to the second prospect. It seemed best to turn back to Las Vegas. After discussing pros and cons, we decided that it would be foolhardy to return by way of Coal

Valley and its mud-filled dry lake. It would take no longer to return by way of Pioche, the roads were better, and there was no chance of getting mired again. There was less chance also of having a rear end tear itself to pieces. Anyway, Pete said, if he was late in reaching Carson City, the men he was to meet there would wait.

We pulled out of Adaven for Sunnyside as the first light of day was showing. On our way Pete decided that we would not stop to look at the other prospect but go on to Pioche and from there to Las Vegas. When we reached Pioche, Pete was going to send a telegram to the men waiting for him in Carson City saying that he would be a day or two late in arriving. Pioche was the only place on our route from which a message could be telephoned or telegraphed.

We arrived in Sunnyside in time to have breakfast with Whipple and his cowhands. Whipple owned the largest ranch there, and both Pete and I knew him well. During breakfast the storm that had been threatening for several days let go in earnest. Neither Pete nor I cared to make a drive to Pioche in such weather. We remained at Whipple's the following two days and nights until it was over.

The storm had not laid much snow on the ground, at least at Whipple's, and when we started out we believed that we would have no trouble making Pioche before noon, a distance of about fifty miles. The seven miles of road before we started west over the pass across the Ely Range was level, and there was not enough snow on it to prevent the dirt of the road from being seen, for wind had blown most of the snow away.

As we went up the road to cross the divide, the snow became deeper and in places had drifted into snowbanks several feet deep, which we had to shovel our way through to keep going. The farther we went the worse the going became, until we could make no progress whatever without first cutting boughs from cedar and pinon and laying them on the road for us to pass over.

Making progress was time-consuming, but we were getting along, although several times we were on the verge of turning around to go back to Whipple's. A few hundred feet from the crest of the divide, the Studebaker's rear end gave a roar and cracked up again. There was no decision for us to make—we could not walk to Pioche, and we had no alternative but to return to Whipple's on foot.

It was late in the afternoon, and for a time we debated whether to sleep in the car over night or to plow our way on foot through the snow to Whipple's. The storm had started up again, and we decided to go back down the road we had just driven, where we would have ruts in the snow to follow, which would be easier walking over now and which would probably be filled by drift snow in the morning.

Darkness came quickly, and the snow was already filling the ruts the car had made. Although walking for a time was helped by the boughs we had cut to put on the snow so that the car could move ahead, by the time darkness came there were no ruts left, and no boughs that we could see to walk on. The only way we could follow the road was by going through openings in the bushes, which were silhouetted against the snow.

It was not long before we did not try to find the road any longer and made our way as best as we could through openings between the cedars and pinon, all of the time continuing to go down hill. By the time we had covered three miles and were near exhaustion, we heard the sound of bells and knew there was a sheep camp nearby. We could not understand why sheep would be in the high ranges in the winter season and for a time, until the sound of bells became louder, believed we might be losing our minds.

As we stumbled on through the hard falling snow toward the sound of the bells, we began to hear the bleating of sheep and we knew then that a sheepherder's camp was nearby. The herder's wagon was warm, and we shared his bedding with him until morning came.

During the night it had stopped snowing, and after breakfast was eaten the sheepherder offered to drive us to Whipple's. We were still tired from our ordeal of the night before, but because the herder wanted to get the sheep out of the snow into the valley of the winter range as quickly as he could, we refused his offer and told him we would walk.

We floundered through snow until we reached the road we had driven over the day before. We started down it toward Whipple's. We had walked less than a mile when we saw a horse-drawn, covered buckboard coming up the grade, and behind it two cars. We stopped under a pinon where the ground was almost clear of snow and waited.

When the convoy reached us, Whipple was driving the buckboard ahead of the cars to break trail for them; on the seat beside him was the sheriff of White Pine County from Ely, and in the cars a posse of seven deputies whom he had brought with him. All were looking for Pete to rescue him from whatever trouble he was in.

The Sheriff of White Pine County and his posse had started out at the urgent request of the sheriff of Clark County, who had telephoned him from Las Vegas. When Pete had not returned to his home in Las Vegas within the five days he had said that he would, his wife became anxious for his safety. She knew he would not be late for his appointment in Carson City unless some disaster had befallen him, and she had telephoned the sheriff.

Pete and I got in the buckboard with Whipple and the sheriff, and we drove up the grade to the Studebaker. We had not been at the Studebaker more than five minutes when, from over the ridge from the opposite direction, came another posse, headed by seven deputies riding horses and making trail for the car that followed, in which was the sheriff of Lincoln County. The sheriff had organized his posse in Pioche the night before, also at the request of the sheriff of Clark County, and started at daylight. His posse had to come across the mountains while the White Pine posse came down the White River Valley, where the road was a graded, surfaced road.

Before we had time to transfer our equipment into the cars, another posse arrived from Las Vegas. The sheriff, who was in the lead car, had started with six cars, but arrived with three and his twelve deputies. The three other cars were mired in the mud of the Coal Valley dry lake.

We transferred our equipment from the Studebaker into the car the Clark County sheriff was driving, left the Studebaker where it was, and headed back for Las Vegas. Pete was four days late in keeping his appointment in Carson City. When I returned from China years later, I saw what remained of the Studebaker still sitting where Pete and I had left it. Pete didn't want anything more to do with it after he drove away from it that day, nor did anyone else.

The two years that followed my return to the Boss Mine at

Goodsprings after John T and I had taken the trip with C. W. and Nevius to San Francisco and Nevada City in 1915, were busy ones, and mostly routine. Sometimes, however, routine was rudely shattered by some unexpected event, such as the Goodsprings flood in the late summer of 1916.

Goodsprings was built on a rise of ground sheltered on three sides by low hills. The eastern side was unsheltered and sloped into the bed of a mile-wide dry wash that had its head in the southern end of Spring Mountain near Charleston Peak, thirty miles to the northeast. Rain had deluged the town and surrounding hills and mountains in fitful but local cloudbursts for days.

One evening all hell broke loose, and the clouds let everything in them go while lightning lit the landscape for many miles around. The storm continued into the morning and added tons of water to an already over-wetted land. The mile-wide dry wash east of town was running deep with water, and every dry wash, gulch, and canyon was pouring thousands of gallons more into it by the minute.

The Goodsprings line, which boasted three houses and twenty girls, was built on a low ridge east of town in the wide dry wash, and around it was raging water, increasing in depth and madness by the minute. On the crest of the ridge, which had become an island, were the girls and two score men, waving and yelling frantically for rescue. We could hear them above the roar of the water as we stood on the town side, although it was impossible to distinguish what they said.

Rescue over the two hundred feet of raging torrent was possible only with an improvised breeches-buoy, which we rigged up but failed to get into working order; so the marooned girls and their men visitors from town remained stranded that day until the flood water subsided.

When the flood water subsided, the marooned girls and their visitors made their way to safety by holding on to the ropes, which we had got across, and wading almost waist deep through not too quiet water. The girls came first and were welcomed by cheers from the men who were standing by to watch. They were a sad-looking, wet, and bedraggled lot, but the women of town took them to their homes, gave them dry clothes, warmed them, and fed them. At least two of the women who helped the girls had, I knew, once been on the line until they married into respectability. It did not matter though to the others. They were treated as would have been anyone else in trouble.

After the girls came their male guests. Unlike the girls, they were welcomed with ribald comments and a few cat-calls. The wives of those caught on the island were not opposed to the line girls, but they did not want their men playing around where they were, and greeted their husbands coolly. The reason that there were so many men on the line that night was that one of the madames was throwing a birthday party with everything free. Among the men who had been invited, and accepted the invitation to the party, were the elite of the mining and business men of town. It was my good luck, that the invitation sent to me was not delivered until the day after the flood.

It was at the time that I was arranging to get into the war with the Golden Engineer Company of the Colorado National Guard that I received an urgent telegraphic request from Tom Campbell, governor of Arizona, to come to Phoenix immediately. I dropped everything and was on my way to Phoenix the following day. When I walked into Tom's office in the State Capitol (I had telegraphed him the time I would arrive), I found Ted and five others besides Tom waiting for me. Tom did not introduce any of us, although I am certain that most of the men in the room knew each other, although I knew none of them except Tom and Ted.

Tom told the group that he was calling on us to assist in putting an end to the serious labor troubles that had developed throughout Arizona, particularly at the copper mines, although he thought they would spread to the sheep and cattle ranchers. Bisbee and Douglas were the copper mining camps that were the most affected, but he anticipated more trouble at Globe, Jerome, Ajo, and others of the copper-mining centers.

The Bisbee, Douglas, and other copper-mining camp troubles, Tom told us, had no legitimate foundation and had no relation whatever with legitimate labor union grievances between the miners' unions and the operating companies. The troubles were being fathered by German agents in the United States and also followers of the Bolsheviks who were in control of Russia and were endeavoring to spread their communist theories throughout the world.

Tom told us that he had made investigations that revealed that hundreds of out-of-state workers had gone into the copper-mining camps of Arizona and had taken over the branches of the miners' unions, and also that the I. W. W. and Western Federation of Miners were under the control of the agitators. A committee of five known agitators was in complete command, and members of the unions had no say or vote as to what would be done. The rank and file of the unions were loyal men, and were opposed to what was going on and were doing everything they possibly could to stop the trouble, without results.

Appeals had been made to the United States government for armed forces to guard the copper mines and prevent more troubles that would certainly close down the copper mines and seriously interfere with our war effort. The federal government had refused the request, and as a result the difficulties had increased ten-fold.

Tom was a Republican and he told us that although he admired President Wilson in many ways, the President had expounded such liberal views that he could not be trusted to realize events as facts that were in opposition to his social theories, and that the President interpreted events according to some of his academic theories. In addition, Tom said, he was aware that the President had several advisors, in whom he placed utmost confidence, who were sympathetic to the German and particularly to the newly-developed Russian philosophy. Tom did not trust any of these.

Politics were playing a strong part in anything that Tom could do to help correct the situation or to put a stop to the activities of the subversives behind the strike troubles. Should he come into the open and make a fight, certain officials or advisors in the United States government who were close to President Wilson would influence him to act adversely to the interests of Arizona, with resulting danger of disaster to the entire United States. It was for these reasons Tom told us that he was calling on us to give any assistance that we could.

Tom was emphatic in telling us that Sheriff Harry C. Wheeler of Cochise County, where the greatest of the trouble was taking place, and General John C. Greenway had plans formed to meet further trouble that was expected, and that they had the Bisbee and Douglas situation under control.

Not long after the meeting with Tom, the irregular "strikers" were "deported" in box and cattle cars and shipped under guard to the United States military authorities in Columbus, New Mexico. The greatest threat to the copper mines was removed by the deportations, but it was followed by attempts to close the copper mines in other camps in Arizona. Deportation on a lesser scale than those at Bisbee followed, and the legitimate members of the miners' unions weeded out the wreckers and agitators who had come into Arizona for the single purpose of destroying its great copper-mining industry and sabotaging the war effort of the United States.

Soon after the Bisbee deportations I left Arizona and went east to complete those things that I had to do before getting into the war. I reurned to California when everything was in order to join the Golden National Guard Company at Camp Kearney, where it was in training before its departure for France. I did not go with it, for the Chemical Warfare Service asked for my transfer to it, and I went to Cleveland where I designed furnaces to manufacture activated gas-mask carbon so that the men I had wanted to go overseas with might live through poison gas attacks.

John Lamoigne leaving the Furnace Creek Ranch in Death Valley with his four burros, after resting two weeks and replenishing his food supplies. About October, 1910. Photo courtesy Pacific Coast Borax Company, Furnace Creek Ranch, from the collection of Mr. Harry Gower.

Water tank installed by John Lamoigne at Garlic Spring. The wooden tank (right of John's elevated tank) was built by U.S. Army Engineers to impound water for use on field maneuvers. John's first shack stood below the elevated tank; the "castle" was built in the shadowed area between the tank and the left side of the picture, which was taken in 1952. Crampton Collection.

Furnace Creek Ranch, at mouth of Furnace Creek Wash, Death Valley, November, 1919. Crampton Collection.

Our quarters at Furnace Creek Ranch. Under trees such as these the old-time prospectors and desert rats rested at the ranch. They also cleaned up in the water of the ranch ditches while waiting for supplies from Death Valley Junction. Crampton Collection.

"Road" into Death Valley through the Emigrant Wash, August, 1919. This is typical of the "roads" on the desert and into Death Valley as late as 1919. Old John's lead mine is located several miles to the south (left) beyond the ridge. Crampton Collection.

Steve Larson (left) and Jay Ginn at Old John's Garlic Spring camp, before construction of the castle. April, 1917. Crampton Collection.

John Lamoigne and Johnny as they were found by Shorty Harris and the author, August, 1919, nine miles north of Furnace Creek Ranch in Death Valley. Johnny lies in the middle foreground, and Old John's body lies a little to the left and above, under the shade of the mesquite bush at the extreme left. Crampton Collection.

The grave of John Lamoigne in Death Valley, August, 1919. The grave marker was carved at the time of the burial by Harry Gower of the Pacific Coast Borax Company after Shorty Harris and the author had found the body of Old John. The mesquite under which the body was found is at the right, just out of the picture. Crampton Collection.

Remington Hill placer mine, Nevada County, California, operated by the author from 1930 to 1941. From left: John T, holding gold-pan with considerable gold in it; Jimmy James; Elmer Golden with rifle; two miners; Vance Newman with rifle; the author. John T, who had been invited to the mid-summer clean-up of 1935, died soon after this picture was taken. Crampton Collection.

La Brea oil seepage pools near Los Angeles, early December, 1905, taken the day Charlie Waring and the author discovered fossil bones there. The location is now in Hancock Park, on Wilshire Boulevard, Los Angeles. Crampton Collection.

LEGEND OF JOHN LAMOIGNE

Had not the events that followed been memorable, I probably could not now recall the day that Old John made his first and last visit to Piute. It was typical of many other August days on the desert, and the day itself was therefore not unusual. For Old John, however, to drift so far from his regular beaten trails and prospecting grounds was an event that one could not ignore. How he happened to hit the trail to the desert around Piute no one ever knew. Many times I tried to blast and muck it out of Old John, but he would dodge an answer or ignore the question.

It was lunch time, and Ma Gilson had just rung the triangle bell to bring her boarders to lunch in what she called the "Desert Eating Place." She would not call it "restaurant," and I guess "eating place" was more appropriate anyway. Adjoining the eating place, with a door between, was a small, country-like store, and a post office over which Pa Gilson presided. In addition, Pa Gilson had a profitable sideline—he not too surreptitiously sold hard, red, home-distilled, corn whiskey. Prohibition days were many years in the future, and undreamed of, but a sideline of good old corn was not unusual at desert "watering places" such as Ma and Pa Gilson had. "Corn" was dependable for it worked faster than "store" whiskey anyway. After all, it was the results that counted, and Pa Gilson's supply never ran low.

The corrugated iron roof that covered the eating house and adjoining store creaked in the sun like a string of exploding firecrackers. The roof was so hot that its heat, radiated to the room below, made it an oven. Butter on the eating table, melted almost to the consistency of water, had been put into pitchers to be poured. The usually intolerable flies, seeking release through window openings covered with dust-ridden wire screen, fought momentarily to get through to open air, then dropped to the sill below to lie quiet.

Outside, in the shade of a sheet-iron roof that sheltered hard-packed dirt, which took the place of a wooden veranda, the thermometer registered nearly 120 degrees. It was hotter inside the "eating room" and breath-stifling. There was not a breath of wind stirring, although in the distance, occasionally, one could see rat-tail twisters, thin and narrow, rising into the sky, like miniature tornadoes, which they were. The twisters moved across the blistering sands, and, leaving a dust cloud in their wake, disappeared as suddenly as they had appeared.

As the last singing note on the triangle bell sounded, Ma Gilson called, as she had at countless meal times before, "come and get it," and signs of life appeared on the dirt veranda. The old timers who had been sitting on the long board benches began to file listlessly through the screened doorway to the eating room. Ma Gilson's "come and get it" was the last warning. She did not tolerate late comers, for there was more important work to do than feed these desert rats. All of them were good cooks, and they could have dished up food for themselves had not the insufferable heat sapped their will to move, much less to work. Ma Gilson's patience was worn thin on that hot day, and none of those old timers dared risk her wrath. Although motion was slow, it was motion.

Only a few had passed through the doorway to make their way slowly to the tables when a strange noise was heard from the direction of the west road. Looking in the direction of the sound, we saw a huge billow of dust coming from over the ridge and along the road leading to Gilson's. The men who were still outside, and even Ma Gilson, waited to see who and what was coming. No one with any sense traveled this part of the desert at this time of day. Ma Gilson wanted no newcomer anyway. She had counted noses on the veranda and prepared food for them only; another set-up only meant more work.

Obviously it was not a twister, because metallic sounds came from within the rolling billows of dark brown dust that clung close to the ground. Neither could it be one of the Model T's that infrequently made passages into this forbidding part of the desert. A Model T would have its nose out in front, with the billowing cloud following and partly enveloping it. Somewhere within the dust cloud, however, must be a slow moving vehicle. Pa Gilson made the only comment—one that would have been echoed by all, had there been the will to speak, "I wonder who the damn fool is."

As the billowing dust cloud slowly moved down the road toward us, the cause of it reached the dust-swept hard-pan, and Old John with his easily recognized buckboard emerged from the choking dusty mass. The desert rats who had stood outside to watch, their questions answered, filed listlessly into the eating room. Old John could take care of himself, and Ma Gilson's food was tempting even when the weather was insufferably hot.

As I paused in the doorway to look back at the noise and dust, I saw four decrepit, dejected-looking, moth-eaten burros with Old John sitting on a seat behind, urging them to great effort in their slow progress over the dusty, desert road. All were covered and sweat-caked with brown dust. Only the eyes of Old John and the burros could be distinguished as features of living things.

The buckboard was held together mostly by bailing wire tightly bound here and there. From the sides hung a rare and curious assortment of pots, pans, and cooking utensils. On the floor of the buckboard, fastened down with ropes, was Old John's tarp-covered bedroll which he used as the driver's seat. Behind it were a rope-wound windlass, ore bucket, Irish buggy, picks, steel, muck-sticks, single and double jacks, a ladder, and the miscellaneous tools and equipment of a desert prospector.

Old John drove slowly up to the almost shady side of the eating house to a shed under which Pa Gilson kept a few bales of hay for such emergencies as some wandering desert rat whose burros he knew would need food. Under the shed there was also a small watering trough that could be filled by pumping from a not too shallow well. Old John unharnessed and chain-hobbled his burros, gave them each a flake of hay, and from somewhere beneath the ill-assorted buckboard load, brought forth a tomato-can measure of oats for each of the animals.

With his burros cared for, Old John came into the eating house, sat down, and helped himself to food which Ma Gilson had placed on the table family style. With the exception of Ma Gilson's "Pitch in, John, but don't dare be late again," no words had been spoken excepting "Hello, John" and an answering "Morning Bill," "Morning, Steve." That was all. In such weather even this little conversation was difficult. Why talk, anyway? Everybody knew Old John and he knew them, and there were no new strikes to talk about, so why bother?

I had first met Old John in the early Goldfield days. I had come up from Los Angeles on the Mojave-Randsburg-Panamint route to Death Valley, which it crossed on corduroy across the bottom sand dunes of the valley, then climbed its way to Goldfield and Tonopah. The second night out, I stopped at the old Panamint silver mining camp at Ballarat. It was early spring, and every desert rat with a grubstake was on the move; and it seemed that almost all of them were at Ballarat. All were looking for another Goldfield, and each was sure he would be the one to find it. Shorty Harris almost did, and started a boom at Bullfrog; but the gold gave out, and so did Shorty's boom camp.

That evening, sitting before a fire drinking local corn whiskey and swapping stories, Shorty Harris, who was also on his way to Goldfield, came up with a tale of an old prospector who had a camp off the Wild Rose Canyon road near the camp of Skidoo.

Shorty brought out a specimen of silver ore to back up his story. He said that Ole Nelson, a dependable old desert rat, had brought it in. Ole told good lies only when he had poured down too much corn, and when Shorty guaranteed that Ole was sober at the time that he told about this silver prospect, I knew it must be true. Ole had somehow missed a trail and stumbled onto this claim. He had stopped and taken a look at the prospect that he said was owned by a "crazy old Frenchman" named John Lamoigne. There was no question about the ore's being high grade, and Shorty and I agreed this was as good a time as any to go to take a look at the hole. The prospect was only a few miles off the road to Goldfield. We would lose no more than a day of travel anyway, and it might be something really worth looking at.

We had no trouble finding Old John's claims. After we reached his cabin, we followed a trail over a low ridge and looked down over the mouth of a tunnel just as Old John was coming out pushing his Irish buggy loaded with ore. We "helloed," and Old John "helloed" back and told us to come to take a look at what he had just brought out. In a few moments we were standing by Old John and looking at some of the nicest silver-glance ore I have ever seen. After the usual comments, and questions on how much it assayed, Old John asked us to come into the tunnel and take a look at the face.

Inside, after Old John had brought candles and holders for us, we saw a smooth-walled vein about a foot wide that could be mined clean by stripping. It was so high grade it could be sacked without sorting. Old John treated us, as we did him, as though we had known each other for years, and not as the total strangers that we were. That was the way of the desert in those days. Desert rats recognized each other although they had never met before. No tenderfoot could have gone into Old John's tunnel except over his dead body.

After we had gone over the outcrop, looked at the prospect holes Old John had put down, and inspected the sacked ore, we went back to Old John's cabin. After supper there, we shacked up for the night. The next morning we went on our way to Goldfield, leaving Old John busy mining high grade for his next shipment.

It is not certain when Old John first came into Death Valley, but it was probably in 1884. Isadore Daunet, a Frenchman, as was Jean Lamoigne, had written him in France that he had found borax in "the hottest valley this side of hell" and invited Jean to join him to make his fortune; but before Jean reached Death Valley, Daunet had died by his own hand. Old John let me read the letters from Daunet. When Jean arrived in America, he had with him volumes of the classics in French, English, and German, which he kept on shelves in the stone cabin he built below his lead prospect in a canyon west of Emigrant Wash.

Shorty and I never lost track of Old John. He occasionally came into Goldfield, Tonopah, and some of the other boom camps, but Old John just did not like to be around where there were many people, so his visits were short. More often we met either at his claim near Skidoo or in some God-forsaken place where no one would go to find ore, except Old John, who usually found it.

Old John never sold that silver prospect, although he had many offers. It was better than a bank for him. Whenever he needed money, he'd high grade a few tons of ore, and sell it to some local ore-buyer. If he wanted money for more than his usual needs, for a long-time grubstake, or to pay off bills that he owed after a long prospecting trip, Old John would mine enough to make a shipment direct to one of the Salt Lake smelters.

Wherever Old John went, his credit was good. Storekeepers from one end of the desert to the other knew that whenever his bills began to mount up, Old John would go back to his prospect, mine some ore, ship it, and soon afterward come in with cash to pay up.

Old John must have prospected every hill within more than a hundred miles of Skidoo. He usually ranged between Owens Valley to the Amargosa and Pahrump. He had prospects scattered as far south as Calico and Owl's Hole, east to Pioche and Ely, and north to Bishop and Virginia City. He did not go west so far—the Sierras stopped him. His need for money was never great, and it is doubtful had he needed a large sum regularly that his prospect would have furnished more than he mined from it. When I looked at the hole, a few days after Shorty Harris and I had buried Old John, there was not much left to take out. It may be that Old John was looking for another one, in new stamping grounds, when he drifted into Piute. But while Old John lived at his mine, and worked it, his prospect had been dependable and gave him the ore that brought dollars for anything he might need, and at any time he needed them.

As a prospector Old John had no equal. His ways were unorthodox and unheard of in the prospector fraternity. He never looked for anything where anyone else would expect to find it, but where others were afraid to try. There was always something worth digging wherever Old John went, and he found it. When he found good indications, he would open up and prospect the outcrop. If the ore looked good, he put a location notice in a can near his discovery hole, and moved on to look for something else. Old John never built a "discovery" monument or bounded his claims. It was not the monument building work that Old John minded, but the time it took to build them. Monuments just wasted time that he could spend going somewhere or digging at a new prospect.

The hills were dotted with Old John's discoveries and location notices. But he never filed them in a recorder's office. It did not matter that he did not do this. No one would jump Old John's discoveries anyway, although it is doubtful if Old John would have cared had they done so. He had made a discovery, prospected it, and when he found good ore, he was satisfied. His wants were always taken care of by ore from his silver prospect in the hills near Skidoo. Anything else he found was only for the personal satisfaction it gave him. Judging from the things he found, he surely had a lot of fun.

After lunch we tried hard to learn why Old John had drifted south from his regular stamping grounds, where he was going, or what he was going to do, but we received no satisfaction. Later in the afternoon when the sun was getting low, and a mild breeze cooled the desert air a little, Old John harnessed his burros, hitched them to the buckboard, and was off again, headed east.

Several weeks passed before I saw Old John again. Platinum ore had been discovered a few miles north of Goodsprings across the state line in Nevada. The discovery was so unusual, and the ore so high grade, that a boom was on, and the desert rats and tenderfeet were flocking to the new camp of Platena in the usual excited droves. The best road to the new camp, but the least traveled because of its hundred miles of greater distance, passed through Piute. By sheer accident Old John had established himself not many miles east of Piute and within a hundred yards of this fateful road that was suddenly to change the course of his entire life.

Old John had built a nice, one-room shack just below a small spring, usually avoided by those who passed by because of its strong taste and smell of garlic. Some old timer, long before Old John settled there, had posted a sign near the road, "poison water beware," and at the spring itself another sign that gave to it a name: "garlik spring." It was believed that the water contained arsenic, as did water of many desert springs. Old John had torn down the poison water sign, but left the spring name sign where it was. No one ever suffered from drinking the water, though.

The shack Old John had built was of well-seasoned, old lumber, high graded from some abandoned mine or desert rat's home —Heaven alone and Old John knew where. There was a well roofed porch with a board floor. The room inside had a ceiling, and everywhere I looked there was an air of permanency. It was not hard to guess that when Old John started out he had no intention of leaving his own camp near Skidoo for any length of time. But here he was, and settled down as if he were going to make a stay of it.

Just below the shack, and off to one side, Old John had built a shed and corral of mesquite poles and broken boards for his burros, although I doubt very much if Old John ever closed the gate on them. They were at liberty to come and go as they pleased. I noticed, though, that only two burros were grazing nearby, instead of the four I had seen him with last. I also noticed that he had taken the burro harness and hung them under a lean-to he had built on the side of the burro shed. Not only were the harness in the lean-to, which like the burro shed was corrugated-iron covered. but all of Old John's mining tools and other equipment. Something had gone wrong with Old John's plans, that was sure; he would not otherwise have set up such an elaborate establishment and stored his stuff so carefully.

I didn't ask Old John about the harness, but I did about the other burros. He said, "Oh! They'll be back someday. Guess they couldn't take the garlic water." But from the tone of his voice I was sure both of them were out on the desert, where Old John had found them and covered their bodies with rocks. Because Old John's outfit required a four-burro team, and there were only two of his animals left, I suspected that this was the reason he had decided to build himself a shack near the spring. I reasoned, too, that he would probably wait until some wild burros came in for water, trap a couple, and break them to burro harness, then move on. Apparently wild herds had not drifted that way, or if they had, their members had outmaneuvered his trapping scheme. Anyway, he did not move on until long after.

Travel along the once seldom-used road near Old John's cabin had become brisk. The mining boom at Platena, not more than a hundred miles east, had disrupted the desert's peace and quiet, to say nothing of Old John's solitude.

There was very little water in this part of the desert, and every one traveling the road stopped to get some from the spring. It got so bad that Old John even took the "garlik spring" sign down, but even that didn't help any. Where there was a shack, there must be water, so they stopped anyway.

At first the road was mostly used by the desert rats. Old John welcomed these because it gave him news of the new boom camp, and everyone had a chance to swap the new lies born with it. Pretty soon the old timers were out-numbered by tenderfeet who got the idea that desert travel by automobile was as fast and safe as it would be on streets and roads around some city. By the time these tenderfeet had traveled the two hundred miles to where Old John had his cabin, they were more than disillusioned.

Over the Piute road the trip took almost three days instead of the planned-for one. The road, for miles, was a continuous roller coaster. Speed was impossible, and in the desert heat radiators boiled so often that water, carried in small canteens, was soon exhausted, and many walked a lot of miles to get some to carry them farther. Food supplies that the tenderfeet carried, if they carried any at all, were soon exhausted. Gasoline was another unexpected problem; the desert had no service stations in those days. Pa Gilson had gone to the boom camp, expecting to make his, and Ma Gilson had closed the store when the shelves no longer had anything on them. About all that was left at Piute was the post office, and Ma Gilson's eating place, which the tenderfeet could not bring themselves to patronize.

As a result of the tenderfoot travel, Old John found himself selling stuff from his small supply of canned goods, which he providently bought by the case and stored in a little dugout in back of his shack. Old John didn't like the idea of selling supplies under any condition to "these damned tenderfeet," but would give a desert rat anything he had. Desert rats never ran out of anything excepting money, so Old John didn't have a broken pick on them. But his tenderfoot trade was heavy, and Old John's food supply small.

A lot of old timers were going along the road in both directions, so Old John would ask them to get things for him and bring them back from Berdoo, to replace the stuff he had unwillingly sold to the "damn-fool tenderfeet who would die if I didn't sell grub to them."

Old John's store-keeping was something to contemplate. When it came to prospecting, Old John wasn't lazy; he put everything he had into it. But when it came to selling grub or acting as a storekeeper, he just wouldn't move a hand. When finally the tenderfoot trade became more demanding than his normal replacement of supplies would take care of, Old John bought a wide assortment of supplies. He put in more bacon, beans, and flour, and added cases of things he never would use himself. By this time Old John had almost broken a pick with some of the desert rats. His "orders" for purchases in Berdoo took time and crowded off things they wanted to bring back for themselves. They would beef about it, but the stuff always got to Old John for his unwanted customers.

Old John's business had expanded so that he put up a few two-by-twelves along one wall of his shack, with a short one-by-twelve in front to be used as a counter, although he never went behind it when anyone came in to buy things. When one of the unwanted would stop for water, and come in to purchase supplies and to ask what he had, Old John would point to the shelves and say, "There it is. Help yourself." When what was wanted had been put on the counter, the buyer would turn and ask Old John the price. The answer was always the same, "I don't know." This was unsatisfactory, and after repeated I-don't-knows,' it always ended up by Old John's saying "Leave it in that cigar box. Whatever you think is right."

Without exception the transaction resulted in Old John's getting much more than the stuff was worth. He was in a highly profitable store business that would have made Pa Gilson envious. He did not realize it, however, and would not have cared anyway if he had.

Once in a while Old John would grubstake one of the desert rats and think nothing of it. He knew that when any of them ever got a dollar, or struck something, he would get it all back. He did not care whether he got anything from these grubstakes. It amused him a lot, and he had a lot of satisfaction in helping some old timer like himself.

This grubstake business, however, backfired on Old John. One day, two of the desert rats he had grubstaked, Jay Ginn and Steve Larsen, came in for his signature on some papers, which were to sell a half interest in some claims they had located while prospecting on his grubstake. Old John wanted nothing to do with the deal, but Ginn and Larsen insisted. Old John had grubstaked them and that was that: he not only was entitled to a third share, but he had that share. Ginn and Larsen couldn't close the deal without Old John's signing.

There was a lot of argument, there was nothing Old John could do, so he finally signed the papers. The notary Ginn and Larsen had brought along witnessed the transaction, and Old John was given his share of the first payment, which was dumped into the old cigar box that was his cash register. Ginn and Larsen were sure that the joint one-half the three of them were keeping would bless them with a fortune. Old John didn't give a darn about making money, but Ginn and Larsen did.

As it turned out, Ginn and Larsen were right. It was not long after the deal was closed that money began to pour into Old John's hands in such amounts as he never dreamed of, much less wanted. He didn't know what to do with the money. He had no use for it. He did not trust banks. What to do with what he was getting was something he must have sweated over a long time before he had the answer.

The only person Old John could think of whom he knew was honest, and would take care of his money, was Ma Gilson. So he made Ginn and Larsen go down to Piute and bring Ma Gilson back to talk things over. Old John fixed it up with Ma Gilson so that every time a royalty check would come in, she was to go to Berdoo, get cash in gold pieces, and keep them for him. He would not let her take green or yellow backs. They might burn or the trade rats get away with them. The money would have been safer in paper, the trade rats being partial to any shiny object, and Ma Gilson would not have had to carry so much weight from Berdoo to Piute!

The arrangement would have worked out all right for everybody if the money had not come into Old John in such amounts that it worried Ma Gilson. Old John kept insisting that the money must not be put in a bank. Ma Gilson was to keep it in the house, or bury it in some place. Of course it was not long before the story got out as to who was keeping Old John's money. Ma Gilson was scared to death that someone would rob and perhaps kill her. Pa Gilson's being away didn't help any, either, and there were so many tenderfeet traveling the Piute road every day she thought anything might happen. Before the Platena boom, she could have left the gold pieces on the tables of her eating place with no fear, but not after the roads got "cluttered up with polecats" as she would call the tenderfoot travelers.

Things pretty well came to a head one day when Old John sent for Ma Gilson. Ginn and Larsen had brought him a royalty check for an amount that almost made him tear it up, it was so high. Anyway, Ma Gilson came up, took one look, and told Old John it was deep enough—she would not take care of his money any longer. She told Old John to put it in a bank, and if he wouldn't do that, to spend the whole amount. There was a big blow-up. Ma Gilson broke her pick with Old John, and it seemed that he would have to get along without her. Finally, however, Ma Gilson agreed to take the check, but she refused to cash it.

Things ran along this way for a week. Every day Ma Gilson

would go to Old John's, where the two would argue for hours. Each day it had the same ending: nothing could move Old John, or Ma Gilson, either.

It could have continued as a never-ending see-saw battle with Ma Gilson taking Old John's checks but not cashing them, and Old John insisting that she cash them and keep the gold pieces for him. Ma Gilson might have sharpened her broken pick again had not a solution come about in an unexpected way.

I heard the story from her later and wondered how she persuaded Old John to agree to the fool idea. Ma Gilson must have drilled a deep one, loaded it with powder, tamped it hard, put in a darned short fuse, and threatened to spit it if Old John did not agree. Old John was caught any way he looked at it. The round Ma Gilson had drilled and loaded on him was primed so short he couldn't get away. Both would have been blasted either way, so Old John had Ma Gilson put in a longer primer, and took his chance on getting out somehow.

It started when Link Shaw, roving assayer-surveyor, drifted into Piute. Ma Gilson was getting more desperate, and Old John more insistent. Link was honest, and Ma Gilson knew she could trust him. Link had drifted into Goldfield soon after news of its strike got out. His home was Australia, and some lung trouble put him on the desert to clear it up. As an assayer and surveyor he was not much; his jobs only lasted long enough to get him a stake for a couple of weeks until some new unsuspecting sucker picked him up. But Link never lacked ideas, and Ma Gilson knew she would have her problem solved if he thought up something. She was generous with Pa Gilson's corn, and the idea was born.

When Ma Gilson and Link had the idea worked out, they went to Old John's all ready to spit the fuse. Link's idea was for Old John to build a house with every comfort money could buy. A store and a corral with a fitting stable went with it.

It was a hard job to sell the idea to Old John. Link illustrated his plan by sketches of a house the like of which had never been seen in the world before. Old John took one look at the sketches and said he would have nothing to do with Ma Gilson's big idea. Ma Gilson and Link stuck tight to their plan. Ma Gilson wanted to be rid of all that gold and all that might yet come. Link needed a job. The pair of them were drilling a new hole now with double jacks, and the combination couldn't be beaten. Old John did not want anything to do with the house, or the gold, either. Ma's fuse seemed about ready to be spit.

Old John wouldn't agree to anything Ma Gilson and Link suggested. Ma Gilson had guessed that Old John would balk, so she had brought his uncounted gold pieces along with her in several bags. Ma Gilson spit the fuse, and trapped Old John. Ma threatened to walk out and leave the money behind. Old John could take care of himself. Having the money around, Ma Gilson knew, was the last thing that Old John wanted. It was a rotten trick, all right, but Ma Gilson was taking every advantage that she could. There was nothing Old John could do but to say "yes." Ma snuffed the fuse; it could be cut and spit later if anything went wrong with her plans.

In a matter of days the desert peace and quiet was disturbed by an agony of noise. Big trucks from Berdoo were bringing in a vast assortment of building materials, and the sound of workmen's tools and voices filled the air. Ma Gilson had persuaded Old John to accept Link's design, although he felt it would disgrace the desert and every old timer in it.

The atrocity became known as "Old John's Castle."

It was a large, square building, with round turrets at each corner. A church-like spire rose from the middle of a huge flat roof. The porch, surrounding the house (Ma Gilson refused to let it be called a veranda), was fifteen feet wide, with an ornate railing and rococo filigree under its overhanging roof.

Ma Gilson and Link had put in a long shift designing the interior. The first floor had a music room, dining room, kitchen, parlor, and library. Rooms on the second floor were worked into the complexity of a gabled roof with multitudinous dormer windows. Chimneys came up in a dozen places, like quills on a porcupine, from fireplaces in every room.

The house furnishings—rugs, wall paper, curtains, paintings, and the myriad of other things—Ma Gilson had carefully planned for an interior design of her own inspiration. Ma Gilson had the one great, glorious spree of her bleak existence. Like many others, she had longed to buy everything pictured in mail-order catalogs. This was her golden opportunity to use a free, unrestrained, and flowing hand. No catalog or mail-order house was slighted; all received a share of Ma Gilson's bounty.

The library, its shelves built completely around the room between windows, and up to the ceiling, had a magnificent array of hand-tooled, gold-decorated tomes of the classical literature of all the ages. Oak tables and ponderous, uncomfortable chairs made the room one to avoid. The dining room with its black walnut furniture looked like pictures of some medieval baron's castle, with its fireplace fitted with fixings for roasting a small steer. The kitchen was set up with everything anybody could possibly want or, in wildest imagination, ever use. The stove was a dream of some noted hotel chef.

Huge crystal-glass chandeliers hung in every room, and under them coal oil lamps. The wallpapers were alive with brilliantly colored birds and flowers from other lands. The carpets, in red, green, and blue designs, were thick, and any desert rat would have welcomed a piece from any for his bedroll mattress. The parlor had gold ornamental furniture of the best seventeenth-century French tradition.

A water tank in the attic, filled by a gasoline pumping outfit, furnished water to the bedrooms in which baths and wash basins had been installed. There was both hot and cold water, but Old John never used hot water. He was not used to it in the days of his prospecting, and it was too late to start such foolishness now.

The exterior was painted brilliant red with broad white trim. Window sashes and frames were a bilious green, the shutters a light, sky blue. When it was finished, the castle stood apart, visible from a great distance in any direction in which the hills did not obscure the view.

The shack Old John had built for himself and the corral and shed for Johnny were gone. All that remained of Old John's original lay-out was a three-legged water tank below the spring and the store a hundred yards down the road. The five-hundred gallon water tank stood above the ground on three iron legs a short distance below the cottonwoods that marked the spring. The hand pump had been taken away, and a gasoline outfit put in its place.

The stables and corrals for Johnny (Jennie had gone to burro heaven months before) would have held an eight-mule team. The stalls were of quartered oak, and Man of War would have been proud to be stabled there. I passed this marvel many times after it was completed, but saw Johnny near it only on those last few days before the end.

My trips over the Piute road were never more than two weeks apart. I always stopped to talk with Ma and Pa Gilson. Pa Gilson had come back from the Platena boom and did a thriving business on his home-made corn. After leaving the Gilsons, I would drift on to Old John's and trade whatever news each of us had picked up since the last visit.

The castle was no cure for Old John's money troubles, for the royalty checks were getting bigger all the time. Ma Gilson was hounding Old John to put the money in a bank, but he refused. I didn't blame Ma Gilson any. The amount she had in gold pieces was getting larger, and anything could happen with everyone knowing what went on.

Each time I stopped at Old John's, he was less talkative than the time before. He was not himself and something that he would not tell me was heavy on his mind. Then, one morning as I was leaving, instead of the usual "Tap'er light, Frank," he enlarged it to, "She's deep enough, Frank, tap'er light." I called back, "Tap'er light, John. I'll see you in a week." Old John just looked at me and came back, "Maybe, maybe, Frank, but she's deep enough."

When Old John had refused to tend the store, Ma Gilson had taken over the shift. It was not long before Old John made Ma Gilson close the store. He wouldn't let her sell anything, and finally he boarded up the windows and door. Old John wouldn't even let anybody go to the spring. He'd tell them "You damned fools, if you can't get water before you get here, you can die of thirst." He certainly was not himself, and it wasn't long before the desert was alive with talk about Old John's doings.

One day, just about dusk, when I was making my usual stop at Old John's, I noticed he had taken the buckboard from the stable and was pulling it together with bailing wire. I walked over to where he was working and asked what he was doing. It was getting along into the hot days by then, and his answer was surprising. "May want to go somewhere, soon, Frank. She's deep enough. I may want to go somewhere, and I want to have it ready."

I noticed, too, that Johnny was in the corral and the gate closed. Both the repairs to the buckboard and Johnny's new captive status worried me. I knew it wouldn't do any good to ask Old John what it was all about. The answer would be "Frank, she's deep enough."

About this time Old John began buying powder, fuse, and caps, This was not in the pattern of Old John's last few years. He had not prospected since he came to Piute, but it gave a reasonable answer to his strange activities. When Ma Gilson told me that Old John had bought twenty cases of forty, ten rolls of fuse, and only one box of caps, it just didn't sound right, though. He couldn't get a thousand pounds of powder on his buckboard, anyway, and if he did, it would bow like a swayback mule and bust. The only answer to that seemed to be that he was going to look the close-in hills over, just to be doing something again, but twenty cases of powder would last almost any prospector two or three years. In the rush of events at the nearby boom camps nobody paid much attention or thought a great deal about it.

On my last visit at Old John's, he was cheerful. The buckboard was wired up, most of his heavy things were on it and tied down. Johnny was not in the corral, but I noticed the four sets of harness all oiled and polished. Old John was going somewhere —that was certain. When I left the next morning I got the same old "It's deep enough Frank, tap'er light," and then an unexpected "See you soon?" To which I answered, "Next week, John, tap'er light," and I was on my way.

On a brilliant moonlight night, a few days afterwards and almost two years after he first drifted into Piute, there were a series of terrific explosions at Old John's. Everyone knew them to be blasts and that something had gone wrong. Within minutes a great blaze lifted skyward into the still air. No one wasted time in getting under way to see what it was all about.

It was a half hour before the first old timer got to Old John's. The castle, store, stables, and a part of the corral fence were gone. Burning embers were all that were left. Word that something had happened near Piute came through the telegraph operator at Silver Lake railroad station where I had shacked for the night. I hit the trail in a rush and got to Old John's before daybreak. There was nothing that could be done before daylight but look for signs of Old John. There were none.

Morning came and we looked around to see what had happened. Ends of long, burned-out fuse lay on the ground outside the burned area. Across the road from Old John's castle we picked up the trail left by his buckboard, heading north.

We followed the trail until noon, and stopped when it hit miles of sand-blown rock surface. On it no tracks could be seen on the flat sand-etched stones that were characteristic of that area. All of us had hoped Old John would make camp soon after sun-up. It was at least fourteen hours since he had spit the fuses. If old John had not made camp, he would be more than twenty miles ahead of us. There was no way to tell what direction he was headed, either, because of the flat rock surface. Old John knew the desert, and how to get to where he was headed, so we turned back.

While we were trailing Old John, I had noticed that his tracks cut deep into soft ground, much deeper than they should. It was hard-going ground, but the tracks still puzzled me. Old John should be pushing in soft places, but the tracks bit deep into the ground in front.

A war had been fought in the passing years, and I was back home again on the desert. I had been to Goodsprings and other desert haunts. But it had been a tough go and I needed rest; so I headed for Death Valley, the only place where peace and quiet could be found. In crossing to the valley from the Amargosa, a side trip to Greenwater seemed about as good as anything. So up I went. I knew there would be a few desert rats around the old camp and plenty of corn. Two years was a lot of time to be away, and new tales and lies could be listened to and matched. I knew the visit would be worthwhile.

There was nothing wrong with my idea. A lot of old timers were there, all right, and setting themselves up for winter hibernation. Gallons and barrels of corn were in the making, deer-meat was being jerked, and every few days some desert rat would arrive with his burros, or in a Model T, loaded with winter supplies. Steve Larsen, Shorty Harris, Bob English, and a dozen others I knew well were there, and more came in before I pulled out.

In the evenings our thoughts drifted to Tonopah, Goldfield, Rawhide, Shorty's beloved, but busted boom town, Bullfrog, Darwin, Randsburg, Jarbridge, Platena, and a score of others. We had missed none of them. Memories, stimulated by strong, red corn, never failed, as we remembered and told of something long forgotten, lived it again as if years had not spanned the gap between. Whenever the corn got to working well, we made up verses about prospectors and the desert and sang them to "Oh, Susanna."

Many were missing. Old John Lamoigne had never shown up since he blasted his castle. John T was ranching somewhere down

Salinas way, prospecting for lettuce and beans. Jay Ginn had gone on his way; too much corn and prosperity had done him wrong. He should have known better; he was too old. Chas. Waring had gone over a cliff in a snowstorm west of Mojave. Art Clark, youngest of all, signed up with some railroad engineers, and the Germans got him. And so it went. The ranks were thinning fast.

Almost two weeks of this and I was ready to go. Shorty Harris decided to go with me as far as Indian Ranch, at the foot of Wild Rose, in the Panamint. He would shoot some ducks and geese and come back with his Model T filled for a big feed. We stopped a week at Furnace Creek, where we made camp by some willows near the ranch ditch. When we had gotten washed, inside and out, by the welcome soft warm water, we drifted on.

When we had pulled out of Greenwater, little did I know that thirty years would pass before coming to the diggin's again. None of the old timers I had told "Good-bye, tap'er light," that morning would be around. China, Central America, another war, and time were to separate us forever.

Shorty Harris and I buried Old John where we found him, in the heart of Death Valley. He was resting partly sheltered by a scrawny mesquite. A burro collar with traces made from burro harness attached to it was over one shoulder and under his opposite arm. The buckboard was a collapsed mess on the opposite side of the bush from where Old John lay. Old John had been helping pull the buckboard as one should have guessed. The load was too much for a burro team, so he had taken a part of the load himself. Unharnessed, was Johnny tied to the mesquite, lying close to Old John. On the other side of him near the buckboard lay another burro tied to one of its wheels. Old John had trapped a wild one to share the burden with Johnny.

Shorty and I tried to figure it out. It was not easy. There had been no attempt by the burros to get away. The load and the heat must have gotten them all about the same time. It had been August when Old John pulled out from his castle. Death Valley would be a hell in the heat of that month. Probably Old John unharnessed them from the buckboard to let them lie down, then lay down near them to rest under the mesquite. However it was, they never got up again, and the missing evidence of struggle was enough to show that it had not been long.

We buried him in one of the deepest parts of the valley, eight or nine miles north of Furnace Creek Ranch. Harry Gower, who had come down from the ranch with Oscar Denton, Tom Williams, and two Panamint Indians, carved the simple "JOHN LAMOIGNE" on the board placed at the head of the grave. We dug a pit alongside of Old John for Johnny. The new burro we left as we found it, alongside the tongue of the buckboard. At first we were going to bury Old John and Johnny in one hole. We almost did, but decided it would not be right. Johnny was not human, although Old John thought he was.

These old friends should be resting happy. Above, looking down, are the Panamint peaks that hold Old John's silver prospect, near Skidoo. The afternoon shadows from the peaks come down early, and the peace Old John wanted comes with them. Old John and Johnny were going home to the silver prospect, which they never reached. It is just as well, anyway; there was nothing much left for Old John to take out, and he would not have been happy about that.

Not so long ago, I went back to Garlic Spring. Piute had long since passed into oblivion. There were no houses left, and the only evidence that anyone had once lived there were piles of rusted tin cans and level places, brush grown on them, where houses had once stood and from which the lumber from them had long since been high graded by passing prospectors or desert rats. The paved road that passed through the country by-passed where Pa and Ma Gilson's store and the "Desert Eating Place" once stood, by many miles. The old road from over the hill into Piute was no more than a mark, overgrown by brush, its wheel tracks clear because water had used them for drains. The wheels of progress had brought disaster, and time had done the rest—the wheels of progress, countless automobiles, headed east or west, driven by speed fiends in a frenzy of transcontinental travel. For all their beauty and mechanical perfection, not one of those wheels could have made Ma Gilson's desert roads, had their drivers dared to try.

Many years have passed since Old John spit the fuse. There is nothing there but the spring and Old John's five-hundred gallon water tank, leaning on three badly bent legs. A few charred remnants of what were once boards or beams of the castle are not enough to tell the secrets of the past. It is just as well!

Desert holly and cat claw had taken over where the castle and

Old John's first shack had been. Green desert grass still grew from the dampness around the spring. As I looked, I could see Old John, his hands resting on Johnny. He waved, and I waved back. From the stillness I heard Old John's voice "She's deep enough, Frank. See you soon. Tap'er light." I knew that wherever he was, he would be there with the others we knew, and Johnny, tight cinched, waiting and ready to go.

The author about 1909 or 1910 in Arizona, probably in the area of Wickenburg or Date Creek. Crampton Collection.

SONG OF THE DEATH VALLEY PROSPECTORS

These are a few of the verses of the old song composed by prospectors and desert rats of Death Valley which they sung to the tune of "Oh! Susanna."

1. We've roamed the hills and made new trails, our burros by our side;
   We've looked for gold, but ain't found none, Old Timer, don't you cry.

CHORUS

Oh! Oh! you desert rats, don't you cry no more; we've almost reached the Golden Gate, our old pals waiting there.

2. We few are left, the most are gone up on to Heaven's shore;
   And soon we'll be within the gates, Old Timer don't you cry.

CHORUS

3. There's water there, the sun don't burn, the hills are low, not steep;
   And sand don't choke your breath away, Old Timer don't you cry.

CHORUS

4. The wind don't blow, the rain don't rain, the trails ain't got no rocks;
   The weather's mild, just as you like, Old Timer don't you cry.

CHORUS

5. No need to hobble Johnny now, or look for him all day;
   The grass is green and not burned up, Old Timer don't you cry.

CHORUS

6. The steel don't dull; no need to muck, no real work more to do;
   The Lord is there awaiting, too, Old Timer don't you cry.

CHORUS

7. The gulches all are water filled, and gold dust by the pan;
   The next round shot will blast her in, Old Timer don't you cry.

CHORUS

8. So down the drink and make a toast to all who still are here;
   Another one to pals now gone, Old Timer don't you cry.

CHORUS

9. They're waiting there with burros packed, with cinches tight and fast;
   I'll join them now, and wait for you, Old Timer don't you cry.

CHORUS

10. Our pals are waiting patiently, the jacks are reared to go;
I'm going now, good-by old pals, I'll wait up there for you.

CHORUS

11. My last shift's in, I'm on my way, I'll wait until you come;
And start again to look for gold, Old Timer don't you cry.

CHORUS

12. You won't have long to wait no more, maybe a year or two;
And sing again as now we do, Old Timer don't you cry.

CHORUS

13. So tap her light until you come, and when it's deep enough;
Just load her light and tamp her soft, Old Timer don't you cry.

CHORUS

14. And when we all have reached those gates, our pals awaiting there;
We'll roam again on golden trails, Old Timer don't you cry.

CHORUS TO LAST VERSE

Oh! Oh! you desert rats I've landed straight up here,
The boys all say to hurry up, they're waiting here for you.

INDEX